Conference Proceedings of the Society for Experimental Mechanics Series

Series Editor

Kristin B. Zimmerman
Society for Experimental Mechanics, Inc.,
Bethel, USA

The Conference Proceedings of the Society for Experimental Mechanics Series presents early findings and case studies from a wide range of fundamental and applied work across the broad range of fields that comprise Experimental Mechanics. Series volumes follow the principle tracks or focus topics featured in each of the Society's two annual conferences: IMAC, A Conference and Exposition on Structural Dynamics, and the Society's Annual Conference & Exposition and will address critical areas of interest to researchers and design engineers working in all areas of Structural Dynamics, Solid Mechanics and Materials Research.

Alexandra Karlicek · Dagny Beale · Cora Taylor
Editors

Dynamic Environments Testing, Vol. 7

Proceedings of the 43rd IMAC, A Conference and Exposition on Structural Dynamics 2025

River Publishers

Published, sold and distributed by:

River Publishers
Broagervej 10
9260 Gistrup
Denmark

www.riverpublishers.com

ISBN 97887-438-0152-8 (Hardback)

ISBN 97887-438-0164-1 (eBook)

https://doi.org/10.13052/97887-438-0152-8

Conference Proceedings of the Society for Experimental Mechanics

An imprint of River Publishers

Preface

Dynamic Environments Testing represents one of twelve volumes of technical papers presented at the 43rd IMAC, A Conference and Exposition on Structural Dynamics, organized by the Society for Experimental Mechanics, and held February 10-13, 2025. The full proceedings also include volumes on Nonlinear Structures & Systems; Dynamic Substructuring & Transfer Path Analysis; Model Validation and Uncertainty Quantification; Dynamic Substructures; Special Topics in Structural Dynamics & Experimental Techniques; Computer Vision & Laser Vibrometry; Sensors & Instrumentation and Aircraft/Aerospace Testing Techniques; Topics in Modal Analysis & Parameter Identification I&II; Data Science in Engineering; and Structural Health Monitoring & Machine Learning.

Each collection presents early findings from experimental and computational investigations on an important area within Dynamic Environments Testing & other Structural Dynamics areas.

The organizers would like to thank the authors, presenters, session organizers, and session chairs for their participation in this track.

Editor:
Alexandra Karlicek–*Anduril Industries*, MA, USA; Dagny Beale–*Sandia National Laboratories*, NM, USA; Cora Taylor, Michigan Technological University, MI, USA.

Contents

Chapter 1
Comparison of Accelerometer Selection Algorithms for a Modal Filter Application

Tyler F. Schoenherr

Abstract A modal filter is a useful tool to characterize the full-field dynamic response of a structure. It provides the amplitudes of the inherent modes of the structure, which can be used to understand its full-field response. As useful as the modal filter is, the execution of the technique is challenging. Forming a modal filter requires representative modes observed at appropriate measurement locations. Choosing those locations is critical to reducing fitting and over-fitting errors in the modal filter process. One means of choosing the accelerometer locations recently developed is to minimize the Modal Projection Error. This paper examines using the Modal Projection Error and two other means of choosing measurement locations in order to compare their efficacy with respect to the modal filter. This comparison of the different methods are evaluated using an experimental data set and a novel means of comparing the effectiveness of the modal filter.

Keywords Modal · Projection · Error · Filter · Accelerometer

Introduction

One challenging aspect in dynamic environment characterization is the ability to measure the full-field response of a structure. The most direct means of acquiring the response of a structure is to take direct discrete measurements of acceleration using accelerometers. Although data acquisition systems and accelerometers have greatly improved in the past decades, measuring the approximate full-field response of a continuous structure is impractical in most applications.

Strain is caused by relative motion within a structure. Knowledge of the full-field deflection of the structure at any moment of time is imperative if strain and damage is to be quantified in a dynamic environment. Direct discrete measurements are not adequate on their own to calculate the full-field strain in the structure. Acquiring the full-field response requires additional knowledge of the structure in addition to response measurements.

A modal filter is the combination of the structural dynamic properties of a structure in the form of mode shapes and discrete measurements. The mode shapes are curve fit to the measured data through a least-squares process. By combining the mode shapes and measured data, the full-field response of a structure can be calculated.

There are errors generated during the curve fitting process when combining measured acceleration and mode shapes to calculate the full-field response. Such errors have unique challenges in providing meaningful modal filter results and are described in this paper as fitting and over-fitting errors. Intelligent selections of where the discrete measurements are during the dynamic environment can mitigate some of the curve fitting errors.

Several methods with different modal-based objective functions have been previously proposed to improve the modal filter calculation. This paper attempts to compare the efficacy of three existing degree of freedom selection methods: Effec-

This article has been authored by an employee of National Technology & Engineering Solutions of Sandia, LLC under Contract No. DE-NA0003525 with the U.S. Department of Energy (DOE). The employee owns all right, title and interest in and to the article and is solely responsible for its contents. The United States Government retains and the publisher, by accepting the article for publication, acknowledges that the United States Government retains a non-exclusive, paid-up, irrevocable, world-wide license to publish or reproduce the published form of this article or allow others to do so, for United States Government purposes. The DOE will provide public access to these results of federally sponsored research in accordance with the DOE Public Access Plan https://www.energy.gov/downloads/doe-public-access-plan.

Tyler F. Schoenherr
Sandia National Laboratories, P.O. Box 5800 - MS0346, Albuquerque, NM, 87185
e-mail: tfschoe@sandia.gov

Alexandra Karlicek et al. (eds.), *Dynamic Environments Testing, Vol. 7*, Conference Proceedings of the Society for Experimental Mechanics Series, https://doi.org/10.13052/9788743801528-1

tive Independence [1], condition number, and Modal Projection Error [2]. All three of them are examined in this paper to determine if there are benefits of one method over the others and to evaluate how robust the calculations are to changes and errors in the mode shape basis set.

Theory

The focus of this paper is on the modal filter process. The modal filter is rooted in modal superposition theory that states that a structural response is comprised of a linear combination of the structure's mode shapes

$$\ddot{x} \approx \boldsymbol{\phi}\ddot{q}, \tag{1}$$

where $\ddot{x}$ is the vector of physical accelerations at a set of degrees-of-freedom, $\boldsymbol{\phi}$ is the mode shape matrix at the same set of degrees-of-freedom and $\ddot{q}$ is the modal acceleration or amplitude for each corresponding mode. The linear combination of modes in Eq 1 is only an approximation of the acceleration due to modal truncation.

The modal filter is the inverse of Eq. 1 such that the modal accelerations can be calculated with knowledge of the structure's mode shapes and physical accelerations. In order to transform the physical accelerations into modal accelerations, the physical accelerations are projected on to the mode shapes by using the Moore-Penrose pseudo-inverse of ϕ,

$$\boldsymbol{\phi}^{+}\ddot{x} \approx \ddot{q}, \tag{2}$$

where the $^{+}$ superscript denotes the pseudo-inverse of a matrix. The pseudo-inverse in the modal filter is a least squares fitting process for over-determined systems [3]. Even though this paper focuses on the structural dynamics aspect of this process, the mode shape matrix can also be referred to as a basis set of vectors as the vectors do not need to be structural mode shapes [4]. For the remainder of this paper, this matrix is referred to as a basis set.

The least squares fitting process introduces errors in the calculation of the modal coordinates. If the mode shapes in Eq. 2 do not span the space of the physical accelerations, there is a fitting error that is minimized in a least squares sense and the modal accelerations will not combine with the mode shapes to perfectly reproduce the physical accelerations. The second error is an over-fitting of the shapes to the data. This error occurs when two or more shapes linearly combine to fit the physical acceleration that is not fit by the other shapes. This error can end up being detrimental as these subset of shapes will greatly increase in amplitude to either fit noise or fitting error in the basis set.

The over-fitting error is caused when two or more vectors in the basis set can closely reproduce any other vector in the basis set. This allows the least-squares fitting process to inflate all these vectors to fit measurement noise or fitting errors. If the vectors cannot combine to fit errors, then the resultant error from the projection will be a fitting error. It is advantageous to have the error be from the fit as it provides evidence that is easier to interpret and is quantified by the Modal Projection Error (MPE) [5].

One way to reduce the overfitting error is to select a set of degrees of freedom to measure that makes the basis vectors unable to combine to form another basis vector. Many methods of selecting degrees of freedom for ultimate usage in a modal filter application exist. Three of those methods are examined in this paper. The first method is a minimization of the MPE. The modal projection error is a quantity of how well a set of vectors can linearly combine to create a given vector calculated

$$\text{MPE} = \Psi_n^2 = 1 - \bar{\phi}_{Tn}^{+}\boldsymbol{\phi}_S\boldsymbol{\phi}_S^{+}\bar{\phi}_{Tn}, \tag{3}$$

where Ψ_n is the modal projection error of the nth target vector, $\boldsymbol{\phi}_{Tn}$. The set of vectors, $\boldsymbol{\phi}_S$, is fit to the target vector.

The other two objective functions are the maximizing of the Effective Independence (EFI) of the basis set [1] and the minimization of the condition number of the basis set. These quantities do not directly compute the basis vectors ability to linearly combine to recreate a separate vector in the basis set, but these methods have been used in other studies of modal filter applications with success. In order to optimize the MPE, EFI, or condition number, one degree of freedom is removed at a time and the objective function is recalculated. This process is repeated over and over in a Greedy algorithm [6] until a desired number of degrees of freedom are selected.

Test Hardware and Calibration to the Model

In order to demonstrate the efficacy of different instrumentation selection methods, experimental data is ideal. Use of experimental data is ideal because it contains realistic errors in the basis set from the finite element model (FEM) shapes in their ability to perfectly fit the measured experimental data.

The hardware used to generate data for this paper is the frame-plate-wing (FPW) test bed [7]. This hardware is chosen because development of an adequate model is quick and a large experimental data set already exists. A large channel count test provides channel reduction options for the optimization methods. In addition, a large channel count also provides more validation for the basis set of mode shapes. The FPW was tested in the laboratory in a free boundary condition and impacted with a modal hammer. A picture of the test setup can be seen in Figure 1.

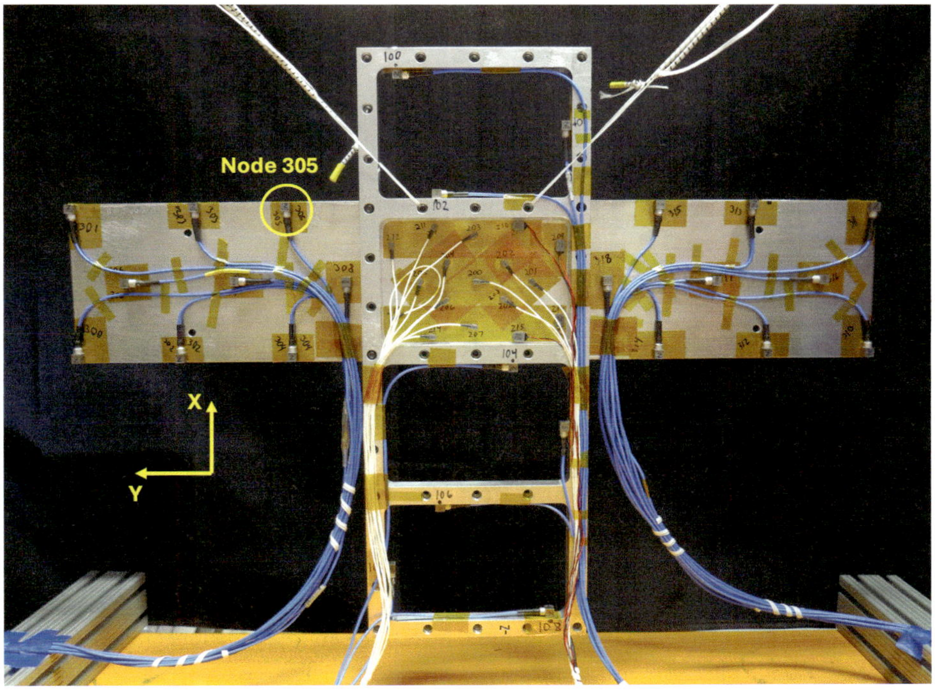

Fig. 1 Photograph of the experimental setup and node locations for the Frame Plate Wing structure

The FEM is compared to experimentally derived mode shapes and natural frequencies. The Modal Assurance Criteria (MAC) with associated natural frequency comparisons can be seen in Figure 2. From the comparison in Figure 2, it can be seen that the model does a good job fitting to the experimental shapes below 800 Hz even though the model natural frequencies are 7-10% higher than the experimentally measured natural frequencies. Above 800 Hz, the FEM shapes begin to represent the physical hardware less. Having modes that match well and match poorly is leveraged in this paper to demonstrate how well the modal filter works with a basis set expected to fit the data well under 800Hz and a basis set with that will incur fitting error under 1800Hz.

Experimental Results

In order to fit a basis set to test data, test data is needed. The response data used in this study is a transient set of data caused by a modal hammer strike of the FPW structure at node 305 in the direction normal to the flat surface of the wing as shown in Figure 1. For all modal filters implemented in this study, the modes are generated by the FEM of the hardware.

The test data is ideal for comparing different means of selecting accelerometer degrees of freedom because the test is highly instrumented at 95 channels. This channel count provides a large pool from which the different algorithms can select degrees of freedom.

Evaluating the efficacy and accuracy of the modal filter is challenging. The output of the modal filter is modal accelerations if the data measured is from accelerometers. Since there is no direct measurement for modal acceleration, success is inferred. With imperfect mode shapes, the modal filter will always be a least squares fit to the test data. Only modes and accelerations measurements from a finite element model would provide a perfect fit and produce exact modal accelerations. Using experimental data, there is always some error in the estimations of the modal accelerations.

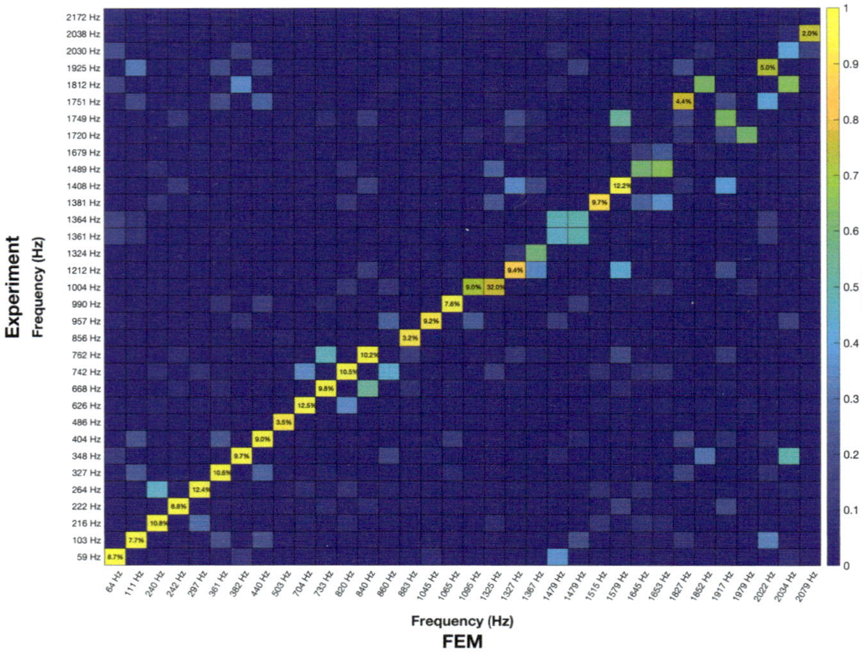

Fig. 2 Modal Assurance Criteria and associated natural frequency error of the Finite Element Model when compared to experimental modal results

It is of interest to alter parameters of the modal filter that have the largest effect on the calculation of the modal accelerations. When analyzing the modal accelerations output by the modal filter, four parameters are considered:

- Method of selecting the degrees of freedom in the filter
- Number of degrees of freedom selected
- Number of modes in the basis set
- Accuracy of the basis set

The changes in modal accelerations when altering the aforementioned parameters are compared to 'truth' modal accelerations. The truth modal accelerations are calculated from using all the degrees of freedom and modes 1 through 32 with modes 20 and 25 omitted in the modal filter. Modes 20 and 25 are omitted because they individually closely resemble other modes in the basis set. Using this set of shapes will be referred to as the full set of shapes.

The frequency content of the truth modal accelerations calculated for modes 1 through 18 can be seen in Figure 3. Because the input force is an impulse from a hammer strike, the rigid body modal acceleration for modes 1 through 6 should be flat at low frequencies with translation in the Z direction (mode 3), rotation about X (mode 4), and rotation about Y (mode 5) being of largest magnitude. These rigid body modes should be highest in magnitude of the rigid body modes due to the location and direction of the force vector. The modal acceleration of the elastic modes 7 through 15 present themselves as single degree-of-freedom oscillators which is expected from modal analysis theory. The modal accelerations in Figure 3 provide evidence that the modal filter worked well enough to use these modal accelerations as 'truth' data.

To increase our confidence in the truth modal accelerations calculated from a modal filter, the modal accelerations are multiplied by the basis set to calculate the response at the measured DOFs . The set of measured DOFs in Figure 4 are not included in the modal filter process so they are considered a blind check of the validity of the modal acceleration calculation. Comparing the response of the measured DOFs to the resynthesized DOFs below 600 Hz, the comparison provides evidence of the modal accelerations being an excellent representation of the response of the system.

To compare the efficacy of the three different DOF selection methods, four modal filters are applied to the test data with varying modal filter parameters. A summary of the modal parameters can be found in Table 1. The first filter uses only modes 1 through 19 and 20 degrees of freedom. Modes 1 through 19 are chosen because they all have high MAC values when compared to the experimental modal data.

The second run also uses modes 1 through 19 in the modal filter, but uses 34 degrees of freedom in the modal filter. The third run uses the full set of shapes and 31 degrees of freedom. The ratio in the third run basis set provides one more degree of freedom than basis vectors. The final run uses the full set of shapes and 45 degrees of freedom.

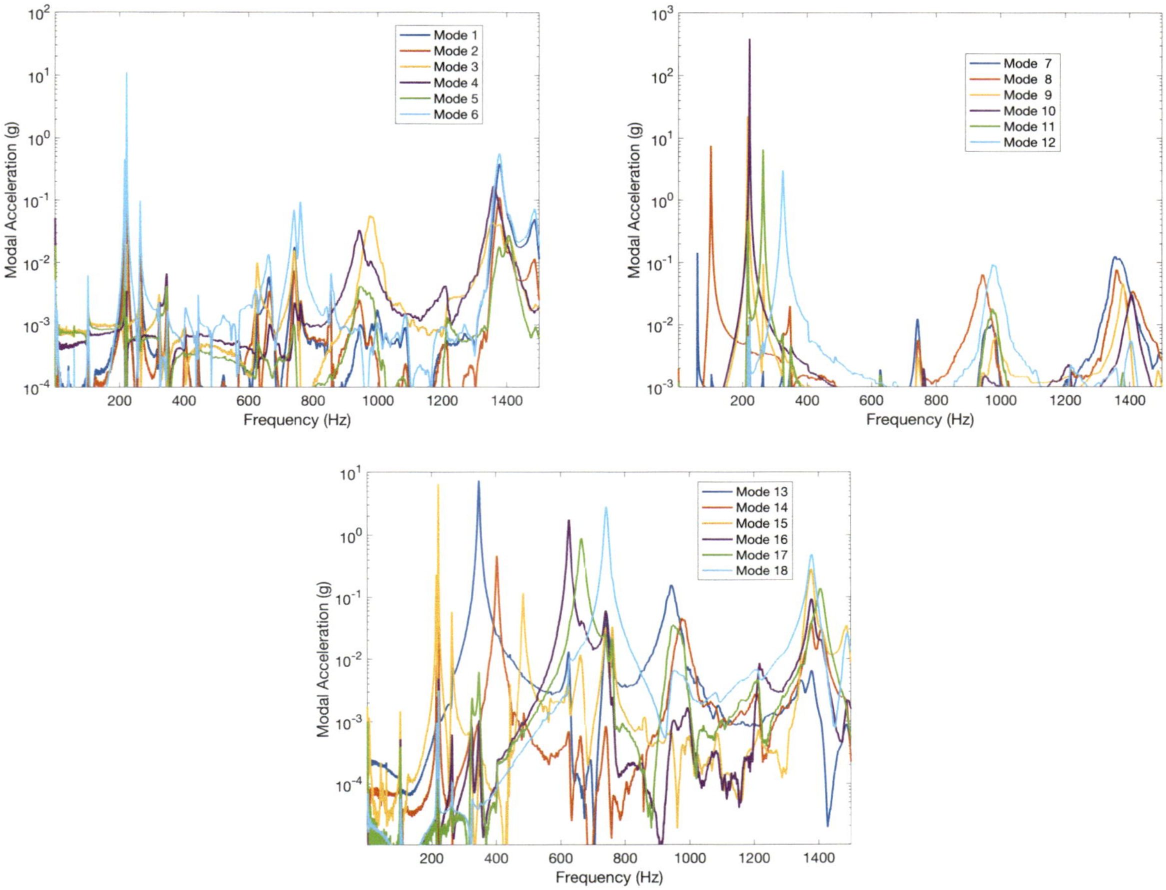

Fig. 3 Modal accelerations of modes 1 through 18 taken to be the 'truth' values

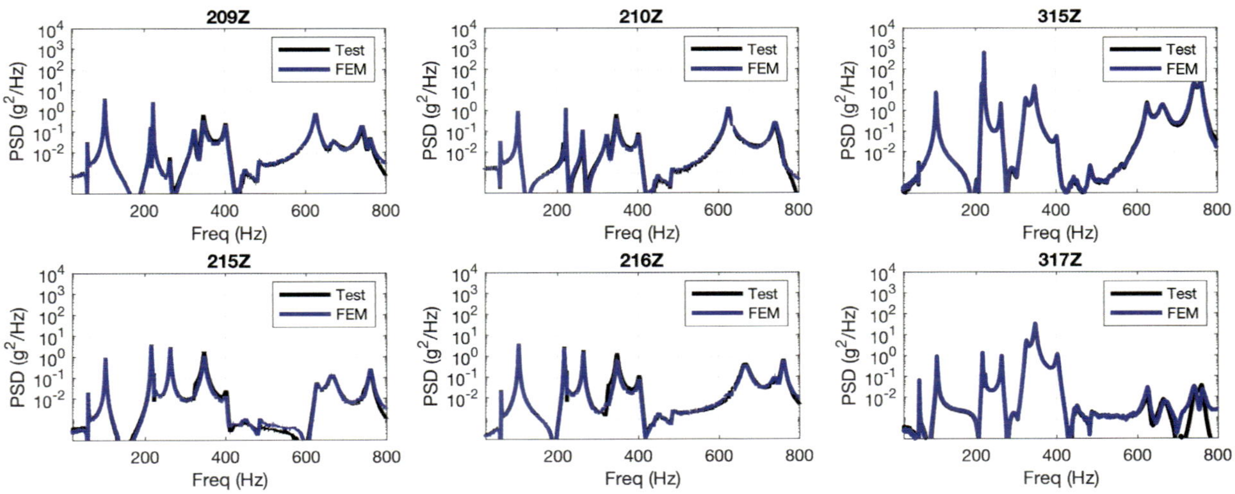

Fig. 4 Selection of degrees of freedoms not included in the modal filter with the modal filter having 19 modes and 20 degrees of freedom in the filter

For all four runs, the different basis sets are used to calculate the modal accelerations. For all comparisons, the peak of the modal acceleration of the Power Spectrum Density is compared to the modal accelerations from the truth calculation.

Table 1 Summary of the four modal filter parameters compared

Run #	Modes Included (m)	# of DOFs in Filter
run 1	1:19	m+1
run 2	1:19	m+15
run 3	1:19, 21:24, 26:32	m+15
run 4	1:19, 21:24, 26:32	m+1

The percent difference to the truth modal peak acceleration is documented and is shown in Figure 5 for comparison. It is important to acknowledge that the real modal acceleration value is unknown, but observing how much the modal acceleration changes when the degrees of freedom of the basis set changes is the focus of this research.

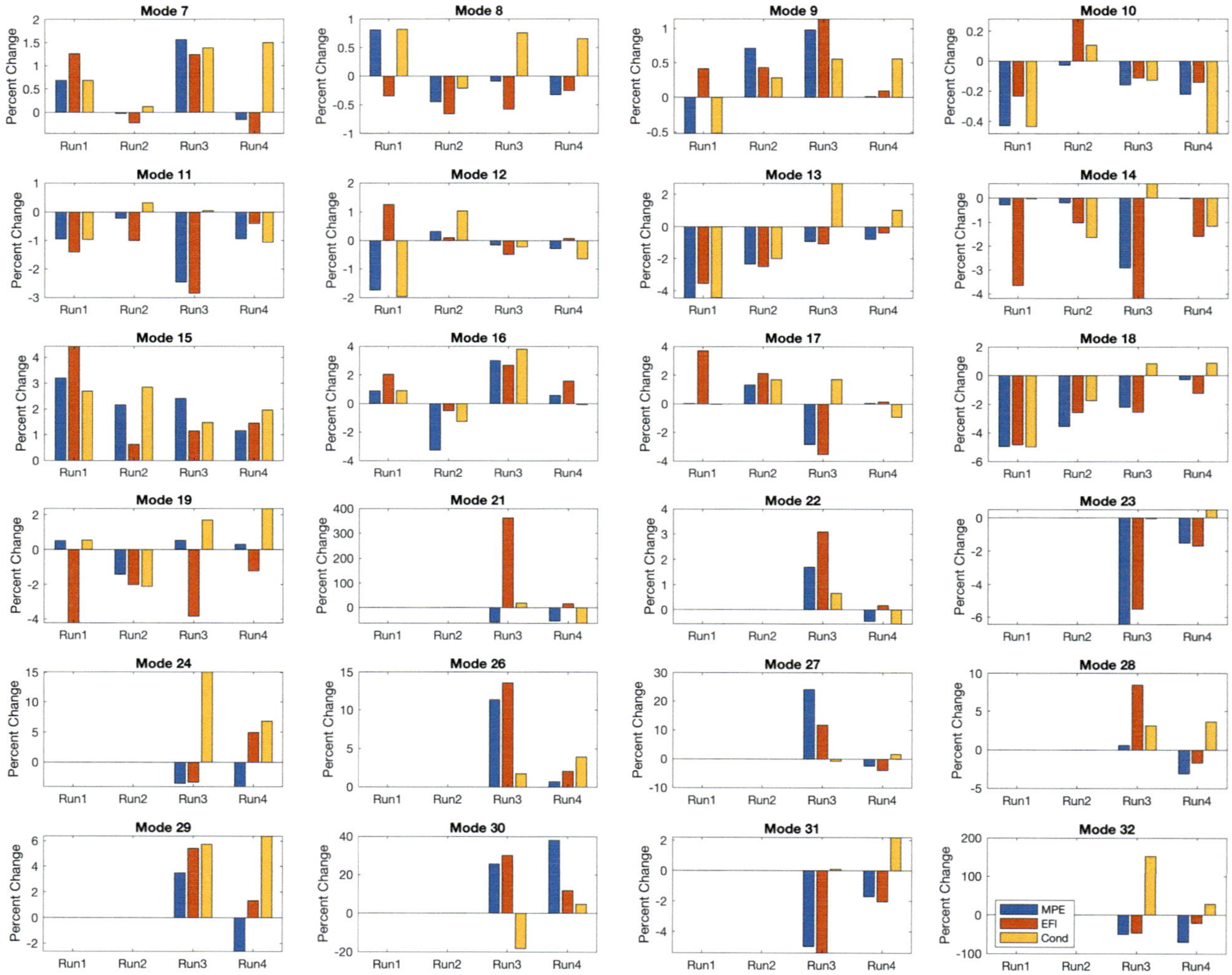

Fig. 5 Error of the modal acceleration at its resonance for each of the FEM modes used in the inverse Run 1: 19 Modes with 20 DOFs in inverse Run 2: 19 Modes with 34 DOFs in inverse Run 3: 30 Modes with 45 DOFs in inverse Run 4: 30 Modes with 31 DOFs in inverse

There are three notable observations of the percent differences of the different selection methods. The first observation is that the largest error occurs for the Effective Independence selection method for mode 21 as the magnitude changes by a factor of 3. However, this result is misleading because the magnitude of the modal acceleration for mode 21 is small enough that it is fitting only electronic noise and not structure response. Small changes to the basis set affecting the modal acceleration calculation is expected when the modal acceleration is in the noise floor.

The second observation is that the methods of selecting DOFs did not appear to have an effect on the stability of the modal filter. Each picking method has modes that caused the result to be the most consistent and the least consistent. Not shown in Figure 5 is the result when a random set of DOFs are chosen. When selecting a random set of DOFs for the basis set identical to run1, the inverse had an average different of around 800%. It is noteworthy that the DOFs that are selected by the three techniques are different from each other. Figure 6 shows the DOFs selected for run 1 by the three methods.

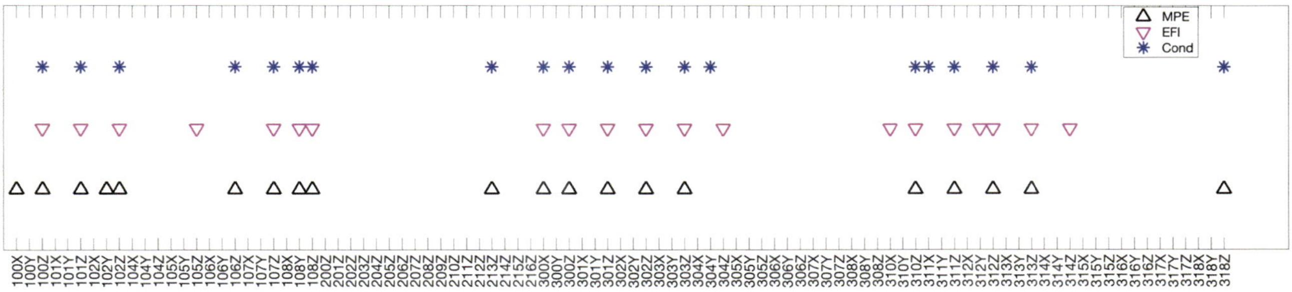

Fig. 6 Degrees of freedom chosen by 3 separate algorithms

The third observation is that the modal accelerations calculated are fairly robust to changes in the basis set given that the basis vectors do a good job of projecting to the measured data. Changes to number of modes included or number of DOFs included generally only altered the magnitude of the modal acceleration by less than 5%. This level of robustness is encouraging for usage of modal accelerations as quantities for characterizing the full-field response in dynamic environments characterization.

Conclusion

This research intended to provide insight into the benefits of using either the EFI, condition number, or MPE DOF selection method over another with respect to a modal filter calculation. In using a relatively accurate model to provide the basis set to fit experimental acceleration measurements, the results show that there is very little benefit of using one selection method over another. However, it is important to use one of the methods as random selection resulted in serious degradation of the modal filter.

The manner in assessing the success of the modal filter in this paper is novel in comparison to other papers on modal filter techniques. This paper uses the modal accelerations to compare techniques instead of the measured and resynthesized acceleration data. Direct comparison of the modal accelerations instead of a subset of measured accelerations provides a full-field response comparison that otherwise is implicitly estimated by comparing selected degrees of freedom.

A benefit to comparing modal accelerations directly is that changes in the magnitude of the modal acceleration is explicit instead of implied when comparing physical coordinates. Additionally, examination of the modal accelerations while altering the parameters of the modal filter provides some information regarding its stability. This is important because an unstable modal filter would be an indication of a poor modal filter. In addition, it appears that most of the modal accelerations change by less than 5%. This conclusion provides a confidence level of using modal accelerations in environment response characterization.

References

1. Daniel C Kammer and L Yao. Enhancement of on-orbit modal identification of large space structures through sensor placement. *Journal of Sound and Vibration*, 171(1):119–139, 1994.
2. Tyler F Schoenherr and Jerry W Rouse. Characterizing dynamic test fixtures through the modal projection error. *Mechanical Systems and Signal Processing*, 204:110746, 2023.
3. Roger Penrose. A generalized inverse for matrices. *Mathematical proceedings of the Cambridge philosophical society*, 51(3):406–413, 1955.
4. Yuanchang Chen. *A non-model based expansion methodology for dynamic characterization*. Thesis, 2019.

5. Tyler F Schoenherr and Jelena Paripovic. *Using Modal Projection Error to Evaluate SEREP Modal Expansion*, pages 111–138. Springer, 2022.
6. Andrew Vince. A framework for the greedy algorithm. *Discrete Applied Mathematics*, 121(1-3):247–260, 2002.
7. D. Roettgen, G. Lopp, A. Jaramillo, and B. Moldenhauer. Experimental substructuring of the dynamic substructures round-robin testbed. In Matthew Allen, Walter D'Ambrogio, and Dan Roettgen, editors, *Dynamic Substructures, Volume 4*, pages 119–123. Springer International Publishing.

Chapter 2
An Improved Fatigue Damage Spectrum for MIMO Random Testing

Enrico Proner and Emiliano Mucchi

Abstract This study focuses on Multiple-Input Multiple-Output (MIMO) vibration control tests, designed to accurately mimic a structure's operational vibration environment. In MIMO random testing, the Cross Spectral Density (CSD) between each pair of control channel plays a key role in the outcome of the test. In fact, the same Power Spectral Densities (PSDs) but with different CSDs may determine a different response on the Device Under Test (DUT) and different drives required to conduct the test. In this scenario, this paper proposes a methodology that aim to replicate a random vibration environment at the component level with minimal excitation. The approach pre-defines the phase and correlation of the reference Spectral Density Matrix (SDM) that maximize the response of the DUT. Then, the reference SDM is completed with the minimum PSD values ensuring that the structure's response in the laboratory aligns with the target vibration environment. This method enhances the consistency of laboratory tests with actual operational environments while minimizing the required energy, ultimately preserving the excitation system's integrity. The paper provides both a theoretical and an experimental analysis of the proposed methodology. The experiments are carried on specifically designed specimen excited by a multi-axis vibration environment, provided by a three-axis shaker.

Keywords Multi-axis test tailoring · Random testing · Fatigue · Fatigue Damage Spectrum · Environmental testing

Introduction

One way to assess the durability of a component is to conduct random vibration tests, which aim to mimic the operational environment by means of random excitation. In general, it is unfeasible to conduct life-long tests, therefore, the vibration environment is usually exaggerated in order to accelerate the test, overcoming time and costs limitations.

Typically, the test profiles are derived from Standards [1, 2]. However, standard profiles are generally much more severe than real vibration environments and may lead to over-testing. The best way to conduct the test is to adopt tailoring techniques [3]. These techniques require to gather operational data to synthesise vibration profiles that are more representative of the operational environment. Nowadays, the most accredited tailoring procedure is the one proposed by Lalanne [4] . Lalanne formalized two types of frequency spectra: the *Extreme Response Spectrum (ERS)* and the *Fatigue Damage Specteum (FDS) [5]*. According to Lalanne, the test article is represented by a of Single-Degree-of Freedom (SDOF) systems, with varying natural frequency in the bandwidth of interest. he vibration environment is applied as excitation to the base of the SDOF system. The FDS for a random excitation can be computed as:

$$FDS(fn) = \frac{n_0^+ T}{C} K^b \left(\sqrt{2} z_{rms}(f_n)\right)^b \Gamma\left(1 + \frac{b}{2}\right) \tag{1}$$

where n_0^+ is the mean number of zero up-crossing, z_{rms} is the root mean square of the SDOF relative displacement and Γ is the gamma function. In particular, the *FDS* represents the damage potential of a vibration environment. Procedures based on the equivalence of the *FDS* are widely studied and researched [6–11]. The formulation of the *FDS* is limited to uni-axial excitations, therefore, it neglects the possibility of multi-axis tests. However, real vibration environments are, in general, multi-axial. Moreover, in the recent past, multi-axis tests have been proven to be more realistic and more severe than traditional single-axis testing procedures [12, 13]. In this context, this paper investigates the possibility of a novel

Enrico Proner · Emiliano Mucchi
Department of Mechanical Engineering, University of Ferrara, Italy, Via Saragat 1
e-mail: enrico.proner@unife.it; emiliano.mucchi@unife.it

Alexandra Karlicek et al. (eds.), *Dynamic Environments Testing, Vol. 7*, Conference Proceedings of the Society for Experimental Mechanics Series, https://doi.org/10.13052/9788743801528-2

formulation of the *FDS*, suitable for multi-axis random excitations. In particular the *Multi-Input FDS (MI-FDS)* is presented. The *MI-FDS* takes into account the CSDs between multiple excitations, allowing to identify the real damage potential of a multi-axis random vibration. Moreover, this work provides the experimental verification of the *MI-FDS*. The experimental verification demonstrates the ability of the *MI-FDS* to evaluate the damage potential of multi-axis vibration environments, which enables the possibility of more accurate multi-axis tests.

The novel proposal: *Multi-Input Fatigue Damage Spectrum*

In this section the novel *MI-FDS* is presented. The general concept of the *MI-FDS*, compared to the traditional *FDS*, is described in Figure 1 .

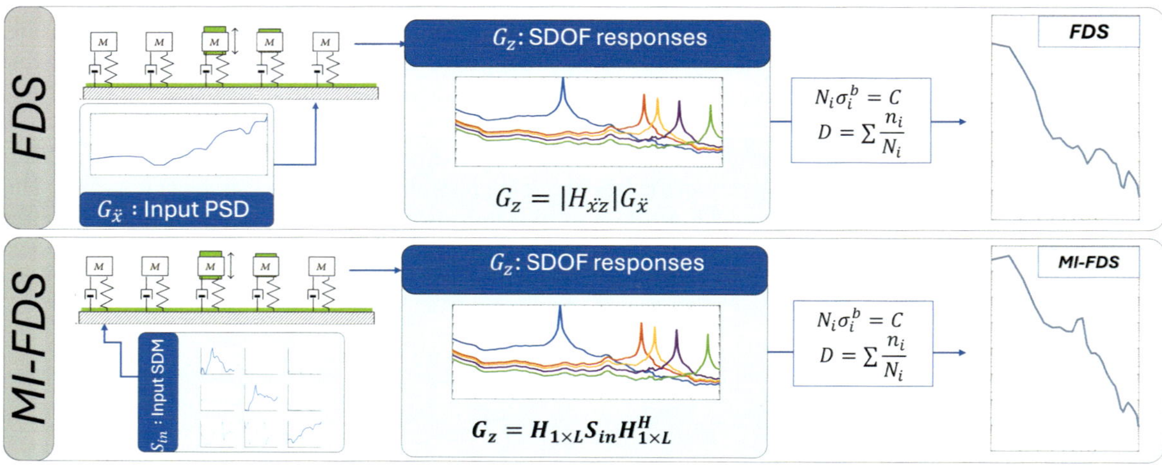

Fig. 1 Comparison between the calculation of the MI-FDS and *FDS* for a random vibration environment.

The *MI-FDS* shares most of the assumptions of the traditional *FDS*. However, the *MI-FDS* considers multiple simultaneous excitations applied to the SDOF system. As a consequence, the total response PSD of the SDOF system G_z subjected to multiple random excitations can be expressed as:

$$G_z = \mathbf{H_{1\times L} S_{in} H^H_{1\times L}} \tag{2}$$

where $\mathbf{H_{1\times L}}$ is the $1 \times L$ transfer function of the system , $\mathbf{S_{in}}$ is the $L \times L$ excitation Spectral Density Matrix (SDM) and the superscript ^{H} indicates the complex conjugate transpose operator. Each element of $\mathbf{H_{1\times L}}$ is computed as a frequency response function with amplitude:

$$|H_{\ddot{x}z}(f, f_n)| = \frac{1}{(2\pi f_n)^2 \sqrt{\left(1 - \left(\frac{f}{f_n}\right)^2\right)^2 + \left(2\zeta \frac{f}{f_n}\right)^2}} \tag{3}$$

and phase:

$$\angle H_{\ddot{x}z}(f, f_n) = atan\left(\frac{2\zeta f f_n}{f_n^2 - f^2}\right) \tag{4}$$

Equation (2) can be also expressed in terms of phases and coherences of $\mathbf{S_{in}}$:

$$G_z = \Sigma_{j=1}^{L} S_{in,jj} W_{jj} + 2\Sigma_{j=1}^{L-1} \Sigma_{k=j+1}^{L} \sqrt{\gamma_{jk}^2 S_{in,jj} S_{in,kk}} |W_{jk}| cos\left(\phi_{jk} - \xi_{jk}\right) \tag{5}$$

where $\mathbf{W} = \mathbf{H^H_{1\times L} H_{1\times L}}$, ξ_{jk} is the phase of Wjk, γ_{jk} is the coherence and ϕ_{jk} is the phase of $S_{in,jk}$. In Equation (5) the phase and coherence of the excitation SDM are explicit. In particular, γ_{jk} and $cos\left(\phi_{jk} - \xi_{jk}\right)$ defines how much the jth and k^{t} excitation are combined constructively or destructively to increase or decrease the response of the SDOF system.

Moreover, it should be noted that the phase difference between $S_{in,jk}$ and W_{jk} is specific of the SDOF system. Therefore, the value $\phi_{jk} - \xi_{jk}$ must be coherent between the test and the SDOF system to guarantee the consistency of the calculated damage potential. The resulting response of Equation (5) can be used to compute the *MI-FDS*:

$$\textit{MI-FDS} = \frac{n_0^+ T \sqrt{2^b}}{C} K^b \left(\int_0^\infty G_z df \right)^b \Gamma\left(1 + \frac{b}{2}\right) \tag{6}$$

Experimental Verification

In this section the experimental verification of the *MI-FDS* is presented. The objective of the test campaign is to demonstrate the capability of the *MI-FDS* to take into account the phase and coherence between multiple simultaneous random excitation and provide an accurate evaluation of the damage potential of multi-axis vibration environments. The first part of this section is dedicated to the presentation of the experimental set up and the results of the testing campaign. In the second part, the results of the experimental campaign will be discussed.

Test set up and results

The specimen chosen for the test campaign is a cantilever aluminum beam with rectangular section and U-shaped notches near the fixed end. A lumped mass of 0.47 kg is placed on the free-end of the beam, to tune its natural frequencies in a convenient frequency range. The detailed geometry of the beam is depicted in Figure 2. The excitation system used for the test campaign is the three-axial electro-dynamic shaker available at University of Ferrara. This excitation system is composed of three actuators (10 kN each) that allow to excite the specimen mounted on the head expander along three orthogonal translational DOFs. Figure 3. SCADAS Mobile SCM202V and MIMO Random of Simcenter Testlab are used as data acquisition hardware and vibration control software, respectively. The control accelerometer (model 356B21 by *PCB Piezotronics*) is placed on the fixture, near the fixed end of the cantilever beam. In addition, an accelerometer is placed on the tip of the beam to monitor the response of the specimen.

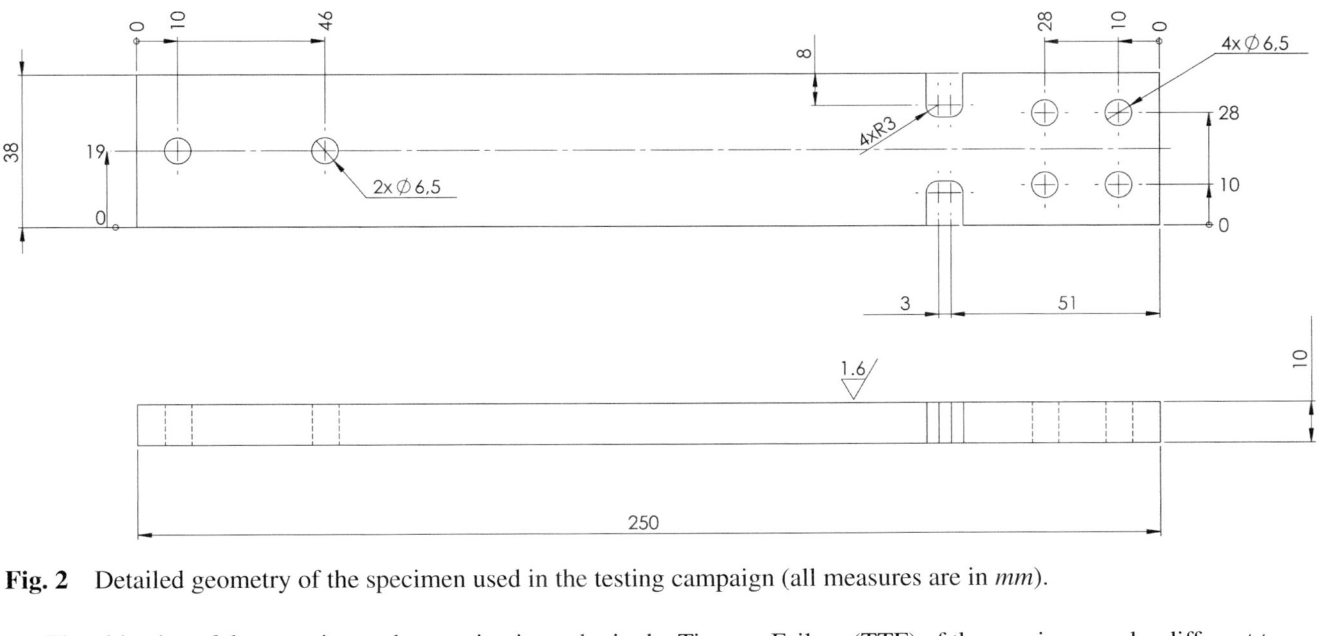

Fig. 2 Detailed geometry of the specimen used in the testing campaign (all measures are in *mm*).

The objective of the experimental campaign is to obtain the Time-to-Failure (TTF) of the specimen under different types of vibration environments. In particular, a reference vibration environment is defined in the $[25 - 200]Hz$ bandwidth. Then, the traditional *FDS* and the novel *MI-FDS* are used to generate equivalent tests. Four kinds of test are carried out, each one repeated on three specimens and conducted until total failure:

1. Test Environment 0 (TE-0): three-axial reference vibration environment.
2. Test Environment 1 (TE-1): synthesise three separate PSDs to have the same *FDS*, computed according to Equation (1), and the same TTF of TE-0. The obtained PSDs are applied separately and sequentially to the specimen for a duration

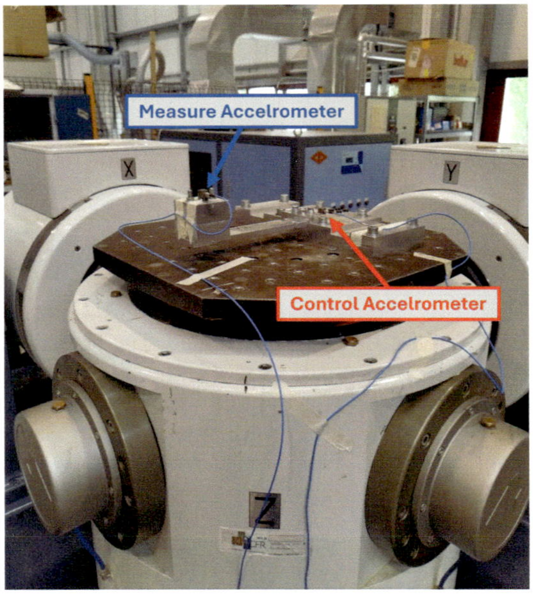

Fig. 3 Specimen mounted on the shaker.

equal to the mean TTF of TE-0. If the specimen survives the first sequence of excitations, the test is performed again until failure.

3. Test Environment 2 (TE-2): the PSDs obtained for TE-1 are applied simultaneously to the specimen. The phase and coherence between each axis is defined according to the *Extreme Dynamic Response Method (EDRM)* [14].
4. Test Environment 3 (TE-3): The vibration environment of TE-2 is scaled to have the same *MI-FDS*, computed according to (6), and the same TTF of TE-0.

The SDM of TE-3 is obtained according to the following equation:

$$\mathbf{S}_{\mathbf{in}_{\mathbf{TE-3}}} = \frac{\textit{MI-FDS}_{TE-0}}{\textit{MI-FDS}_{TE-2}} \mathbf{S}_{\mathbf{in}_{\mathbf{TE-2}}} \tag{7}$$

where $\mathbf{S}_{\mathbf{in}_{\mathbf{TE-3}}}$ and $\mathbf{S}_{\mathbf{in}_{\mathbf{TE-2}}}$ are the excitation SDM of TE-3 and TE-2 and $\textit{MI-FDS}_{TE-0}$ and $\textit{MI-FDS}_{TE-2}$ are the *MI-FDS* of TE-0 and TE-2. The resulting vibration environment is a SDM with the same phase and coherence of TE-2 but with the PSDs that match the *MI-FDS* of TE-0. Figure 4 shows the SDM of TE-0 (solid blue), TE-2 (dashed red) and TE-3 (dashed green). The plots on the diagonal show the excitation PSD along each-axis. The upper triangular plots show the coherence, the lower triangular plots the phase between each pair of excitations.

Figure 5 shows the *FDS* (left plot) and *MI-FDS* (right plot) computed for each test environment. The parameters used for the computation of the *FDS* and the *MI-FDS* are $b = 5.4$, $C = 5.1 \cdot 10^{17}$ and a damping ratio of $\zeta = 0.1\%$. In the plot on the left, the total *FDS* of each test environment is depicted. TE-0, TE-1 and TE-2 are represented by the same curve (dashed gray). The blue curve depicts the *FDS* computed for TE-3. The total *FDSs* are obtained by summing the *FDSs* of each single-axis excitation. The plot on the right shows the *MI-FDS* computed for TE-0 (blue curve), TE-1 (gray curve), TE-2 (red curve) and TE-3 (green curve), computed according to Equation (6).

Finally, Table 1 shows the TTFs obtained for each specimen in the testing campaign. The TTFs from Table 1 are also depicted in the box plot of Figure 6. In Figure 6, each box describes the first quartile of the TTFs obtained from each test environment, the red line and the black cross are the median and the mean of the TTFs, respectively.

Discussion of the results

The first result that needs to be addressed is the difference between the traditional *FDS* and the novel *MI-FDS*. In Figure 5, in the left plot, the total *FDS* computed for TE-0, TE-1 and TE-2 is the same. In fact, the traditional *FDS* considers only the PSD of each excitation. TE-0, TE-1 and TE-2 excite the system with the same PSDs, therefore, the total *FDS* is the same for these test environments. Moreover, the total *FDS* of TE-3 resulted lower than the the total *FDS* of TE-0, as the PSDs of TE-3 are lower. On the other hand, the *MI-FDS* produces three distinct curves for TE-0, TE-1 and TE-2. In particular TE-1

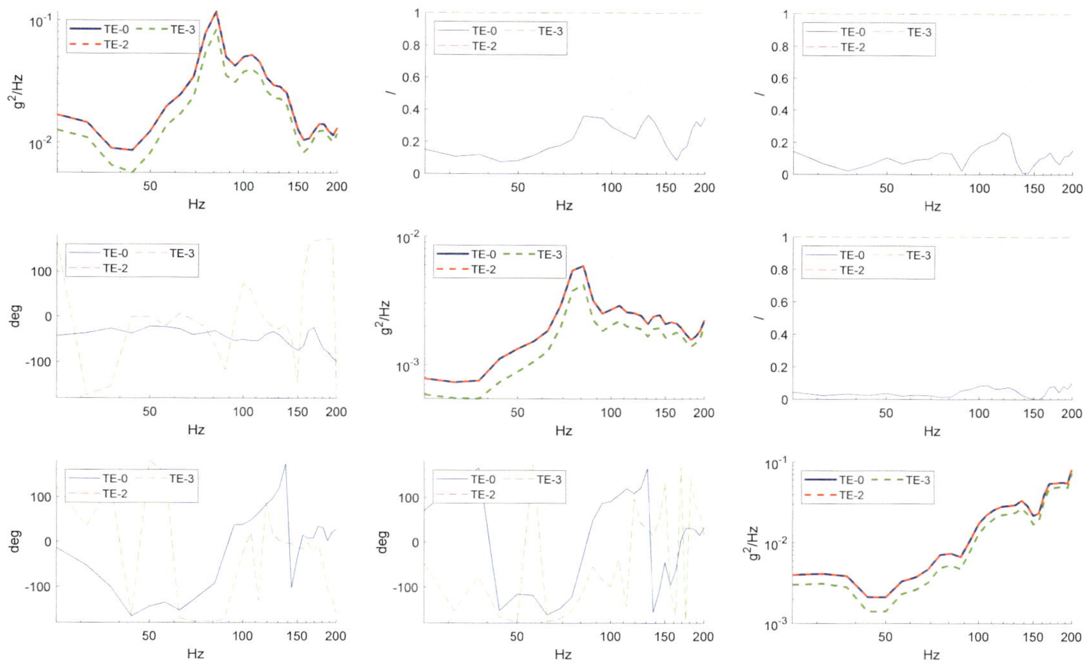

Fig. 4 SDM of the TE-0 (solid blue) ,TE-2 (dashed red) and TE-3 (dashed green)

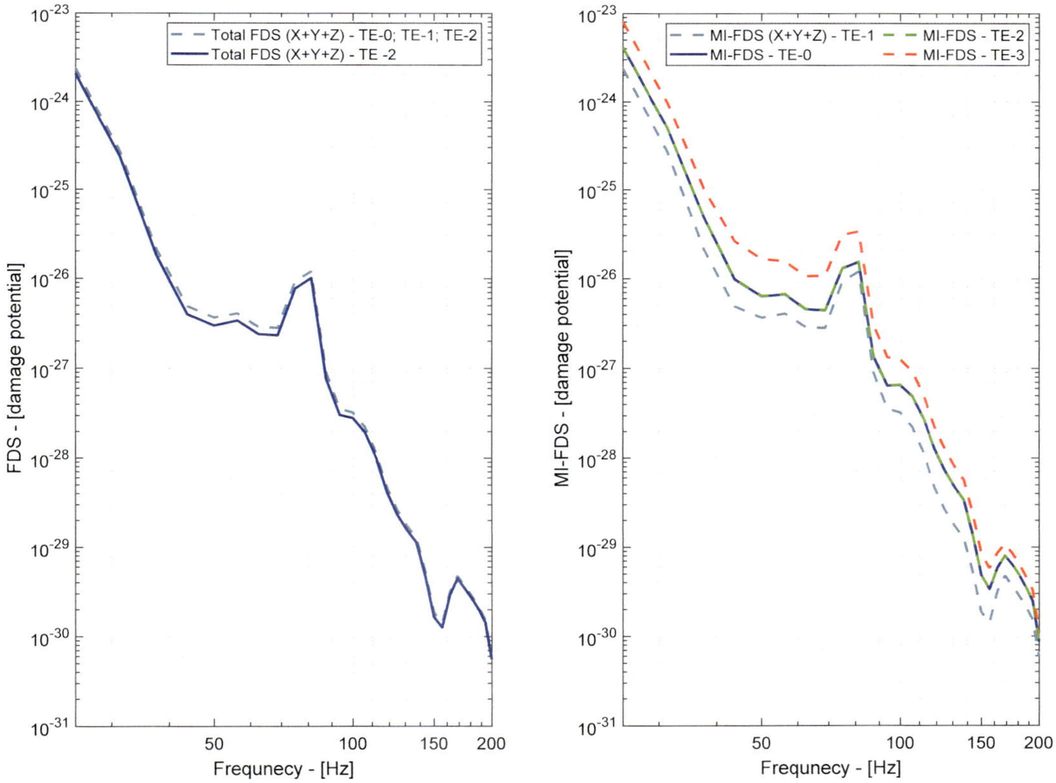

Fig. 5 FDS (left) and MI-FDS (right) of the tests

Table 1 TTFs obtained in the testing campaign

	Time-to-Failure - [s]				
	spec #1	**spec #2**	**spec #3**	**Mean**	**Standard dev.**
TE-0	1537	1620	1920	1692,3	201,4
TE-1	6036	5265	5522	5607,6	392,5
TE-2	1380	1292	1008	1226,6	194,4
TE-3	1648	1818	1472	1646,0	173,0

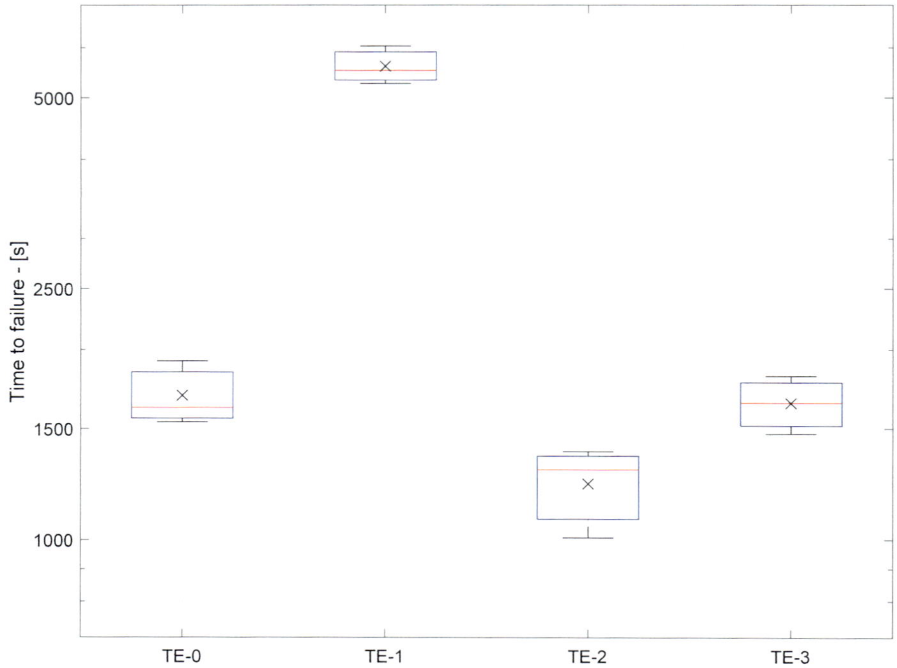

Fig. 6 Box plot of the TTFs obtained in the test campaign

has the lowest damage potential. This is expected, as the combination of multiple simultaneous excitations determines a greater response of the tested system, resulting in a higher damage output [15–17]. The highest damage potential, according to the *MI-FDS*, is given by TE-2. In Equation (6), $cos\,(\phi_{jk} - \xi_{jk})$ and γ_{jk} are both equal to 1 for TE-2. In fact, the phases and coherences of the SDM of TE-2 are set according to the EDRM, which maximize the response of the system and, therefore, achieves the maximum damage potential. Moreover, the *MI-FDS* of TE-0 is lower than the one of TE-2 as the correlation between the excitations determines a lower response of the system. Finally, TE-3 has been obtained to match the *MI-FDS* of TE-0, with the phases and coherences of TE-2. The PSDs of TE-3 are lower than the ones of TE-0, however, the contribution of the CSDs determines an increased response of the system that allows to match the *MI-FDS* of TE-0. The previous consideration imply the following predictions regarding the outcome of the tests:

1. According to the traditional *FDS*: TE-0, TE-1 and TE-2 will be equally damaging to the specimen while TE-3 will be the least damaging environment.
2. According to the novel *MI-FDS*: TE-1 will be the least damaging test, TE-2 will be the most damaging test, TE-0 and TE-3 will deliver the same damage to the specimen.

These predictions are now compared to the actual outcome of the tests, described by Table 1 and Figure 6. The results of the testing campaign demonstrate that the predictions made according to the *MI-FDS* are correct. In fact, in comparison to the mean TTF of TE-0, TE-1 determined a 231% increase in the TTF, TE-2 caused a 27% reduction of the mean TTF and, finally, the mean TTF of TE-3 deviates form the mean TTF of TE-0 by only 2.7%. These results are depicted in Figure 6. It can be seen that TE-1 determined the highest TTFs (least damage), while TE-2 determined the lowest TTFs (most damage), in accordance to the prediction made by the *MI-FDS*. Moreover, TE-0 and TE-3 determined very similar TTFs distributions. This result indicates that, although the PSDs of the TE-3 are lower than the PSDs of TE-0, the damage output of TE-3 is the same of TE-0. This is to be attributed to the contribution of the CSDs, which allow TE-3 to have the same *MI-FDS* of TE-0 but with lower PSDs.

The analysis of the obtained TTFs confirms that the predictions made according to the *MI-FDS* reflect the experimental results of the testing campaign. Therefore, the *MI-FDS* correctly takes into account the contribution of the CSDs of multi-axis random vibration environments to the fatigue damage inflicted to the specimen. Furthermore, the *MI-FDS* has been able to modify TE-2 to generate a new vibration environment that deals the same damage as TE-0 to the specimen in the same amount of time. Finally, the test campaign demonstrates the incapability of the traditional *FDS* approach to evaluate multi-axis vibration environments. In fact, the *FDS* could not distinguish TE-0, TE-1 and TE-2, which have the same PSDs but dealt different damage to the specimen.

Conclusion

In this paper a new approach to the Fatigue Damage Spectrum has been investigated. The *Multi-Input FDS* is proposed, to consider the correlation of multiple simultaneous random excitations in the computation of the damage potential of a multi-axis vibration environment. The validity of the proposed approach has been tested in a dedicated test campaign. In particular, ad-hoc designed specimens have been tested until failure under different vibration environments and the *MI-FDS* has been used to evaluate the damage potential of each vibration environment. The proposed approach allowed to make predictions on the outcome of the tests coherent with the experimental results. In fact, the *MI-FDS* was able to correctly identify the most and least damaging environments and recognised the environments that determine the same damage output. Moreover, the *MI-FDS* allowed to synthesise a new vibration environment equally damaging to the tested article as the reference vibration environment. The experimental results proved that the *MI-FDS* offers a significant improvement over the traditional *FDS* in evaluating the damage potential of multi-axis random vibrations.

References

1. NSA. "Environmental conditions". *AECTP 200, category 240* (2006)
2. de la Defense Delegation Generale pour l'armement, M. "Essais generaux en environment des materials (general tests of materials in environment)". *GAM EG-13.*
3. of Defence, U.S.D. "Environmental engineering considerations and laboratory tests. method 527.1: Multi-exciter test". *MIL-STD 810G w/CHANGE 1.*
4. Lalanne, C. "Specification development, third ed.". *John Wiley & Sons, Inc - ISTE, London* (2014)
5. Lalanne, C. "Maximum response and fatigue damage spectra – part i & part ii". *The journal of environmental sciences, July/Augst* (1984)
6. Xu, F., Ahlin, K., and Wang, B. "Optimization of damage equivalent accelerated test spectrum derivation using multiple non-gaussian vibration data". *Journal of Sensors*, 2021(1):3668726 (2021)
7. Coutinho, I.A. and Landre Jr, J. "Enhanced tailored durability test specification for multiaxial dynamically excited structures attached to a single input base under off-road loading environment". *Engineering Structures*, 219:110768 (2020)
8. Decker, M., Kinscherf, S., Bauer, N., David, P., and Serifsoy, M. "Deriving fatigue equivalent power spectral density spectra for the vibration testing of engine components". *Materialwissenschaft und Werkstofftechnik*, 49(3):392–405 (2018)
9. Cianetti, F., Alvino, A., Bolognini, A., Palmieri, M., and Braccesi, C. "The design of durability tests by fatigue damage spectrum approach". *Fatigue & Fracture of Engineering Materials & Structures*, 41(4):787–796 (2018)
10. Angeli, A., Cornelis, B., and Troncossi, M. "Synthesis of sine-on-random vibration profiles for accelerated life tests based on fatigue damage spectrum equivalence". *Mechanical Systems and Signal Processing*, 103:340–351 (2018)
11. Aimé, M., Banvillet, A., Khalij, L., Pagnacco, E., Chatelet, E., and Dufour, R. "A framework proposal for new multiaxial fatigue damage and extreme response spectra in random vibrations frequency analysis". 213 (2024)
12. Proner, E., Mucchi, E., and Tovo, R. "A relationship between fatigue damage estimation under multi-axis and single-axis random vibration". *Mechanical Systems and Signal Processing*, 215:111402 (2024)
13. Nath, N. and Aglietti, G.S. "Study the effect of tri-axis vibration testing over single-axis vibration testing on a satellite". In *2022 IEEE Aerospace Conference (AERO)*, pages 1–10 (2022)
14. D'Elia, G., Mucchi, E., and Dalpiaz, G. "A novel methodology for dynamic response maximisation in multi-axis accelerated random fatigue testing". 181 (2022)
15. Gregory, D., Bitsie, F., and Smallwood, D. "Comparison of the response of a simple structure to single axis and multiple axis random vibration inputs" (2008)
16. Whiteman, W. and Berman, M. "Inadequacies in uniaxial stress screen vibration testing". *Journal of the IEST*, 44:20–23 (2001)
17. French, R., Handy, R., and Cooper, H. "A comparison of simultaneous and sequential single-axis durability testing" (2006)

Chapter 3
Characterization of Load Uncertainty on Simulated Dynamic Responses

Beale Dagny and Yarberry Kieran

Abstract Finite Element Analysis (FEA) of complex systems such as aerospace structures are routinely being used to predict dynamic responses, generate environmental specifications, and qualify component designs. Uncertainties in model parameters cause model form error, which can lead to overdesign, overtesting, or component failure caused by inaccurate model predictions. Calibration to experimental modal tests address most of these uncertainties. One major source of uncertainty not addressed in system modeling is the uncertainty in the input force. Although accelerations are routinely measured with accelerometers in laboratory and flight testing, the actual input forces being imparted to the system are unknown. Use of the measured accelerations as input forces is known to cause large model form errors. This study characterizes causes of the load mismatch and quantifies the uncertainty in the resulting response predictions caused by the current methods. The Box Assembly, Removable Component (BARC) academic structure is used to perform the load uncertainty characterization. The results of this work are critical to reduce uncertainty in model predictions and enhance the credibility of FEA models.

This article has been authored by an employee of National Technology & Engineering Solutions of Sandia, LLC under Contract No. DE-NA0003525 with the U.S. Department of Energy (DOE). The employee owns all right, title and interest in and to the article and is solely responsible for its contents. The United States Government retains and the publisher, by accepting the article for publication, acknowledges that the United States Government retains a non-exclusive, paid-up, irrevocable, world-wide license to publish or reproduce the published form of this article or allow others to do so, for United States Government purposes. The DOE will provide public access to these results of federally sponsored research in accordance with the DOE Public Access Plan https://www.energy.gov/downloads/doe-public-access-plan

Keywords DET · Force estimation · Modeling · Simulation

Background

Finite Element Analysis (FEA) models are universally used to predict a structure's response to a given excitation. For structural dynamic FEA, those predicted responses will be used to inform component design, supplement the generation of environmental specifications in the absence of test data, and even support component qualification. Uncertainties in model parameters lead to model form errors, which can contribute to overdesign, overtesting, and even component failure. Model calibration is routinely performed to reduce structural model form errors. Model calibration is traditionally performed by iteratively updating the mass, stiffness, and joint definitions in a model until the FEA mode shapes are representative of measured mode shapes of the structure. However, more complex model updating techniques exist that attempt to use advanced algorithms to identify model parameters that will reproduce measured modes, such as FEMtools [1] and FuSED [2].

Regardless of the model calibration approach, a calibrated model is only useful for predicting dynamic responses if the loads applied to the model are realistic. Model calibration is possible because the true dynamics of a structure can be experimentally measured. The true loads on a structure, however, are not possible to measure directly and must be derived from other measurements, usually from acceleration measurements. In some cases load estimation is straightforward; single degree of freedom (SDOF) shaker tests apply a measured excitation to a structure in a single direction. Assuming the shaker is rigid, applying the measured table acceleration as an enforced boundary condition to a calibrated model will accurately predict the structure's response.

Beale Dagny · Yarberry Kieran
Sandia National Laboratories, Albuquerque, NM, 87185
e-mail: djoffre@sandia.gov; dmbeale@sandia.gov; kyarber@sandia.gov

Alexandra Karlicek et al. (eds.), *Dynamic Environments Testing, Vol. 7*, Conference Proceedings of the Society for Experimental Mechanics Series, https://doi.org/10.13052/9788743801528-3

In other cases, however, load estimation is difficult and the error in the estimated load is unknown. For example, a structure experiencing launch conditions will be subject to three-dimensional loads at the attachment locations to the launch vehicle, in addition to possible acoustic loads. Many techniques have been developed to reconstruct impact [3] and even acoustic [4] forces on a structure, but these techniques usually require more measurement locations than are actually captured, and can sometimes generate unrealistic and incorrect predicted loads. More commonly, measured accelerations near the base of the structure are used as applied accelerations in the model. The use of the measured accelerations assumes that the structure between the measurement location and the actual load locations is perfectly rigid, and that the measured location perfectly captures the loads imparted to the structure.

Finally, force reconstruction techniques are useful if the derived forces are reapplied to the structure under the same boundary conditions. In practice, however, forces are derived to define the environment on a specific component within a larger structural system. The boundary conditions of that component are almost never replicated in the lab testing. Reapplying the derived forces to the component with different boundary conditions is expected to produce different responses. However, the error of those responses is unknown.

This paper uses a simple academic structure to quantify how errors in the loads applied to the model can propagate to errors in the predicted response. First, we will quantify how error grows as the load path between the measured base excitation and the actual load location becomes more flexible. Next, we will show that reapplying derived forces to a dynamic structure with different boundary conditions produces different responses, and quantify those differences.

Modeling and Simulation Approach

The Box Assembly, Removable Component (BARC) academic structure was used to quantify how errors in the input forces propagate to predicted responses, and is shown in Figure 1. The BARC was chosen because it is dynamically active, familiar to the Structural Dynamics community, and not overly complex.

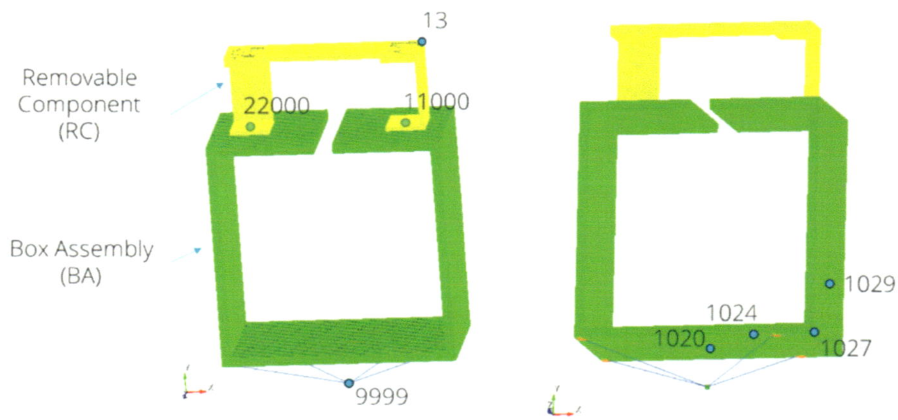

Fig. 1 Box Assembly, Removable Component (BARC) academic structure

The RC connection to the BA was simulated with a full rigid set on the base of each RC leg connected to a small rigid patch on the tops of the BA sides with a stiff spring element for each leg connected to the nodesets 11000 and 22000 shown in Figure 1. The rigid sets are free to move and rotate, but the surfaces in the set must move together rigidly. Each stiff spring element has spring components in all six DOFs. Forces in the spring element can be computed to pull the true force applied to the component at the connection location. The three components of the RC (the top plate and two C-channels) are "bolted" together by tying rigid spiders in the bolt holes together with stiff spring elements. The base of the BA is attached to a concentrated mass (conmass) at four small rigid patches. A conmass was used to represent the structure as if it were bolted to something large and stiff like a shaker table.

The Sierra SD finite element analysis (FEA) code [5] developed at Sandia National Labs was used to simulate the BARC responses. Within Sierra SD, connection forces can be extracted from linear elements, including the spring elements. These forces describe the force imparted at that connection location and will be used as the "truth" forces on the component. The forces can be directly reapplied to the structure. Accelerations, however, cannot be directly applied as defined inputs. Acceleration responses will be converted to force inputs using the equation $F = M*a$, in which a desired acceleration scaled by the mass of a conmass can be applied to a conmass boundary to produce the desired acceleration at that boundary.

Analysis: Reapplied Acceleration Measurements

The first study used accelerations at points systematically farther away from the true excitation location as inputs to the BARC conmass, Nodeset 9999, to quantify how inputs derived from measurements that only approximate the input acceleration will propagate error to the predicted responses. A single point, Nodeset 13, identified in Figure 1 at the top corner of the RC was used to compare responses between reapplied acceleration case studies. Only one point was used to reduce the complexity of the comparison and to represent expected comparisons between test series with limited measurement locations. Four locations were used to represent the measured input acceleration locations, shown in Figure 1. Three input locations were selected on the bottom of the BA: 1020 in the center, 1024 on the edge near the rigid connection to the conmass, and 1027 on the corner. One input location, 1029, was taken near the base on the side of the BA to represent instances where measurement sensors cannot be placed on the base of a structure of interest.

First, a haversine pulse signal with a pulse width of 1e-3 seconds was applied to the conmass of the BARC assembly in the X direction (the direction along the long side of the BA base, shown in Figure 1) and the response of the structure was recorded. The excitation pulse was applied in just one direction to simplify the analysis and to represent single-axis shocks that are common in testing. Next, the simulated accelerations at the four chosen locations were recorded and the X-direction accelerations were converted to the equivalent force to be applied to the BARC conmass. Y and Z accelerations were not included to replicate the case where the direction of excitation is known. The moments were not included, because moments are generally not measured during tests. The acceleration response was predicted using a modal transient simulation with 500 modes of the BARC assembly capturing responses up to 74 kHz, which fully captures the response of the structure in the 2 kHz frequency range of interest. The computed forces from the acceleration measurements at the four chosen locations were then independently applied to the assembly conmass, and the responses of the BARC assembly were recorded. Finally, the maximum percent error of the response at Nodeset 13 was computed.

Figures 2 and 3 show the time and frequency comparisons, respectively, of the acceleration at the conmass boundary condition. Nodesets 1020, 1024, and 1027 on the base of the BA closely replicated the true input pulse, although with some additional inputs that are observed as small perturbations about zero after the initial pulse. Nodeset 1029 on the side of the BA underpredicted the initial pulse acceleration but continued to significantly excite the boundary condition after the true pulse event, which can be seen as peaks in the frequency plot below 2 kHz. The RMS error between the input derived from Nodeset 1029 and the true input was about 30%.

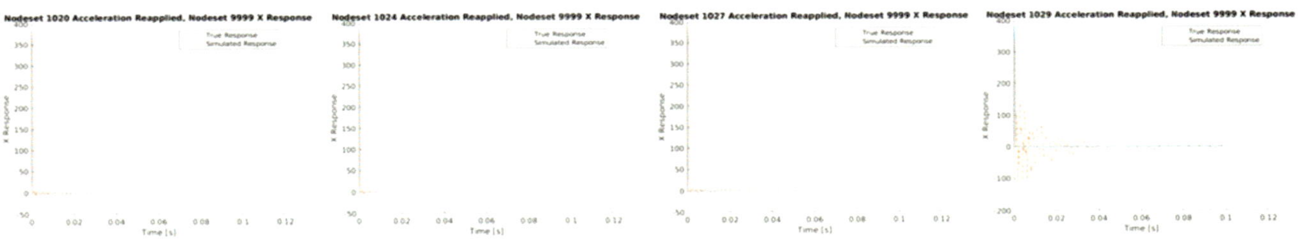

Fig. 2 Acceleration response at forcing location

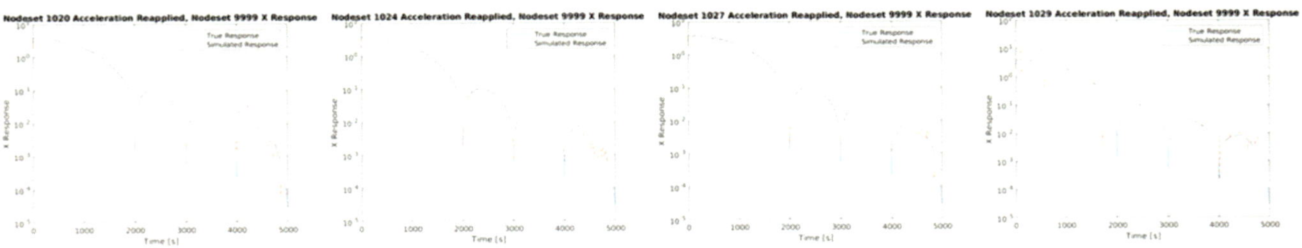

Fig. 3 Acceleration frequency response at forcing location

Figures 4 and 5 illustrate the time and frequency responses, respectively, at Nodeset 13 on the top corner of the RA. Due to the similarities between the true pulse and the applied pulse, responses from inputs derived from nodesets 1020, 1024, and 1027 were very similar to the true responses with rms errors below 5%. The input derived from Nodeset 1029 on the side of the BA, however, produced a response with rms error of 164% in the X direction, 244% in the Y direction, and 23% in the

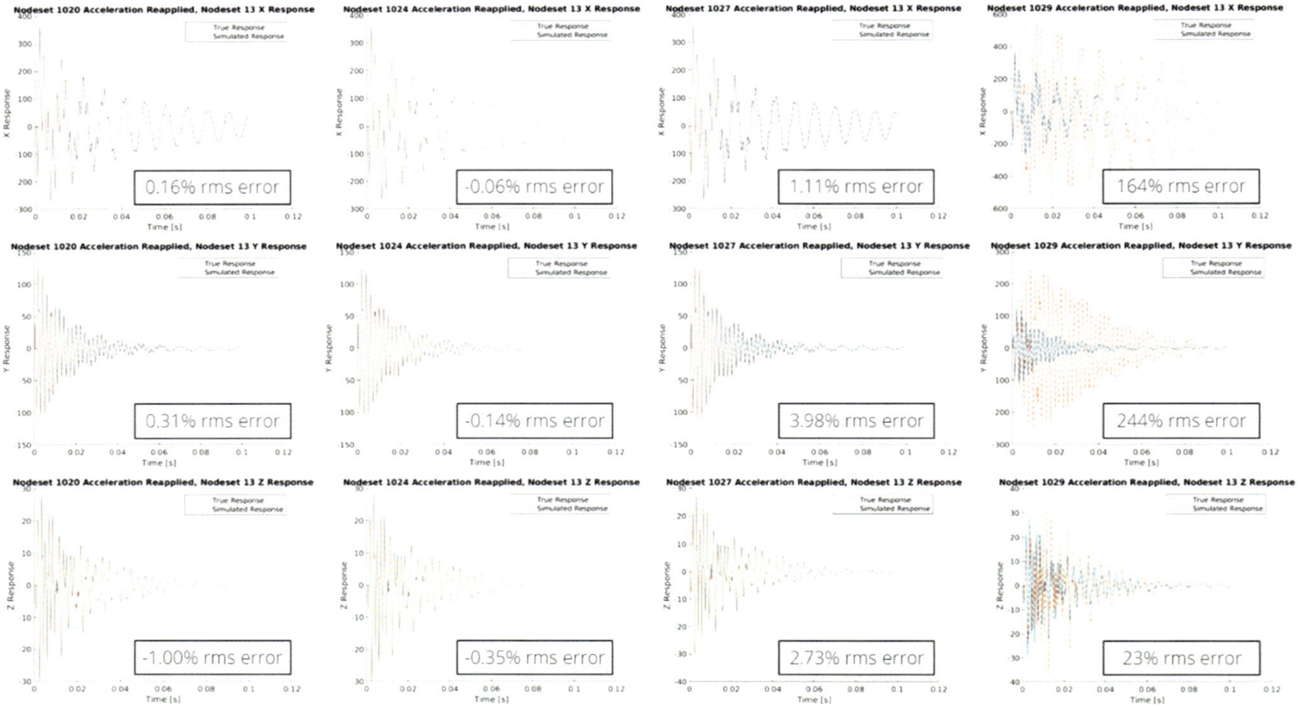

Fig. 4 Reapplied acceleration response comparison at top corner of the RC with rms error noted

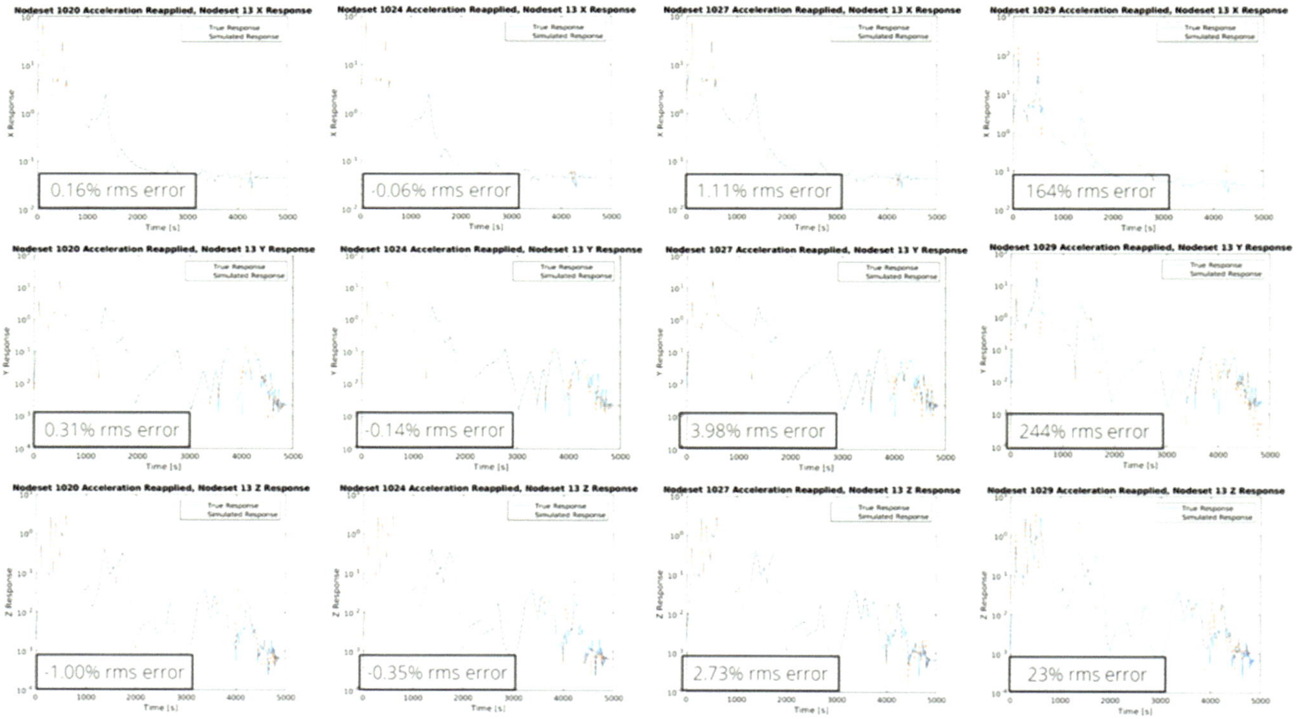

Fig. 5 Reapplied acceleration response frequency comparison at top corner of the RC

Z direction. The predicted response errors were significantly higher than the input error of 30%. The frequency domain plot in the X direction shows that even though the Nodeset 1029 input excitation under-excites at some frequencies, peaks in the input frequency domain translate to peaks in the structural response that overall produce a response that is much higher than the true structural response. In practice, the use of an excitation location like Nodeset 1029 that does not correctly capture the input forces may result in severe overtesting.

This simple exercise demonstrates that errors in inputs can dramatically change the response of a structure, and that the error in input does not predict the error in the response. In this example, three of the chosen input locations were similar enough to the true input to reasonably predict the BARC's response; those inputs were derived from locations that were essentially rigidly connected to the true excitation location. The nodeset that did not correctly capture the true input excitation was dynamically active, and therefore added additional excitation resulting in an overtest of the structure. The nodeset 1029 location used to derive the input excitation was more responsive at structural resonances, shown as peaks in the input spectra in Figure 3, and so the derived excitation also had peaks at those frequencies. Because those peak frequencies align with the resonances of the structure, the derived excitation will strongly excite those modes. Amplification of the structural resonances that were not strongly excited by the true input excitation amplifies the error in the simulated response, which is why the input error is not predictive of the response error.

The chosen location was reasonably close to the true excitation location, and hypothetically would be used to derive inputs to the system if the base measurements were unobtainable. The BARC structure is fairly flexible and the magnitude of the error is magnified by that flexibility. However, in practice, input measurements are often not rigidly connected to the true location of excitation and input error magnification likely occurs.

Analysis: Excitation Force Reapplied

The previous analysis showed that errors in inputs propagate to predicted responses, and that predicted responses can magnify those errors. However, the previous analysis assumes that the true boundary conditions of the system are known. In practice, true boundary conditions are rarely known. A second academic study was performed to demonstrate how the same forces applied to a component in different boundary conditions will produce significantly different responses.

The BARC model was again used, but now with a focus on only the RC component response. The same pulse input was applied to the conmass of the full BARC model and the connection forces between the RC legs and the BA top was computed for the connection location at each leg of the RC. Because there are no other forces applied to the BARC system besides the impulse at the conmass, the connection forces perfectly describe all forces acting on the RC. A hypothetical perfect force reconstruction technique performed on the RC would produce the calculated forces shown in Figure 6. However, reapplying those forces to the RC component alone does not produce the same response at Nodeset 13, as shown in Figure 7. The RC response to the same forces applied at the same locations is different because the boundary conditions have changed; the stiffness and mass at the applied force locations have changed.

The fact that the same forces applied to a system with different boundary conditions produces different responses is known. However, this example illustrates that force reconstruction techniques are limited when the boundary conditions

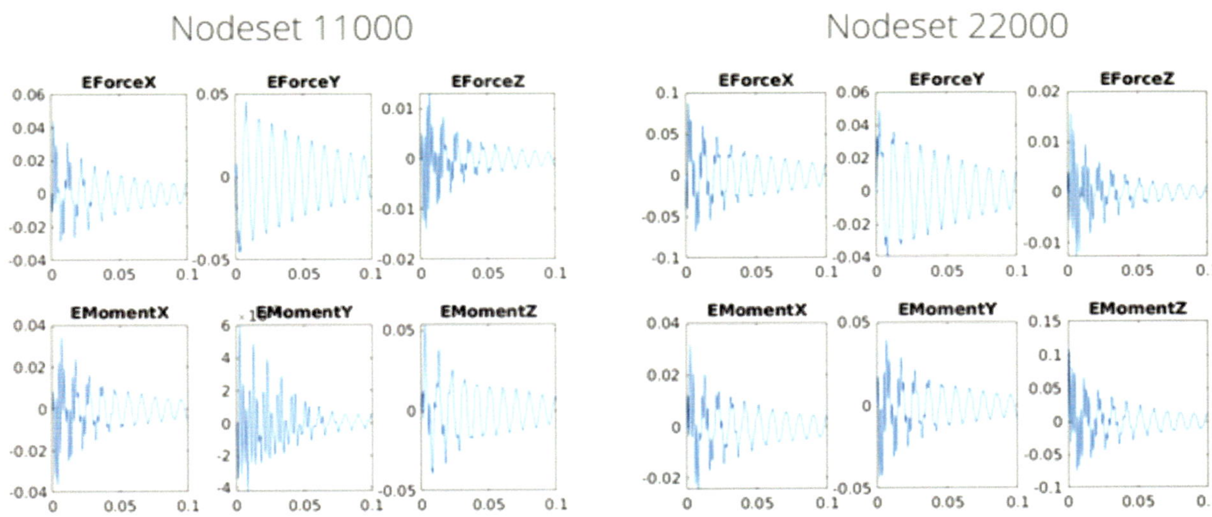

Fig. 6 Applied force computed at the RC leg connections

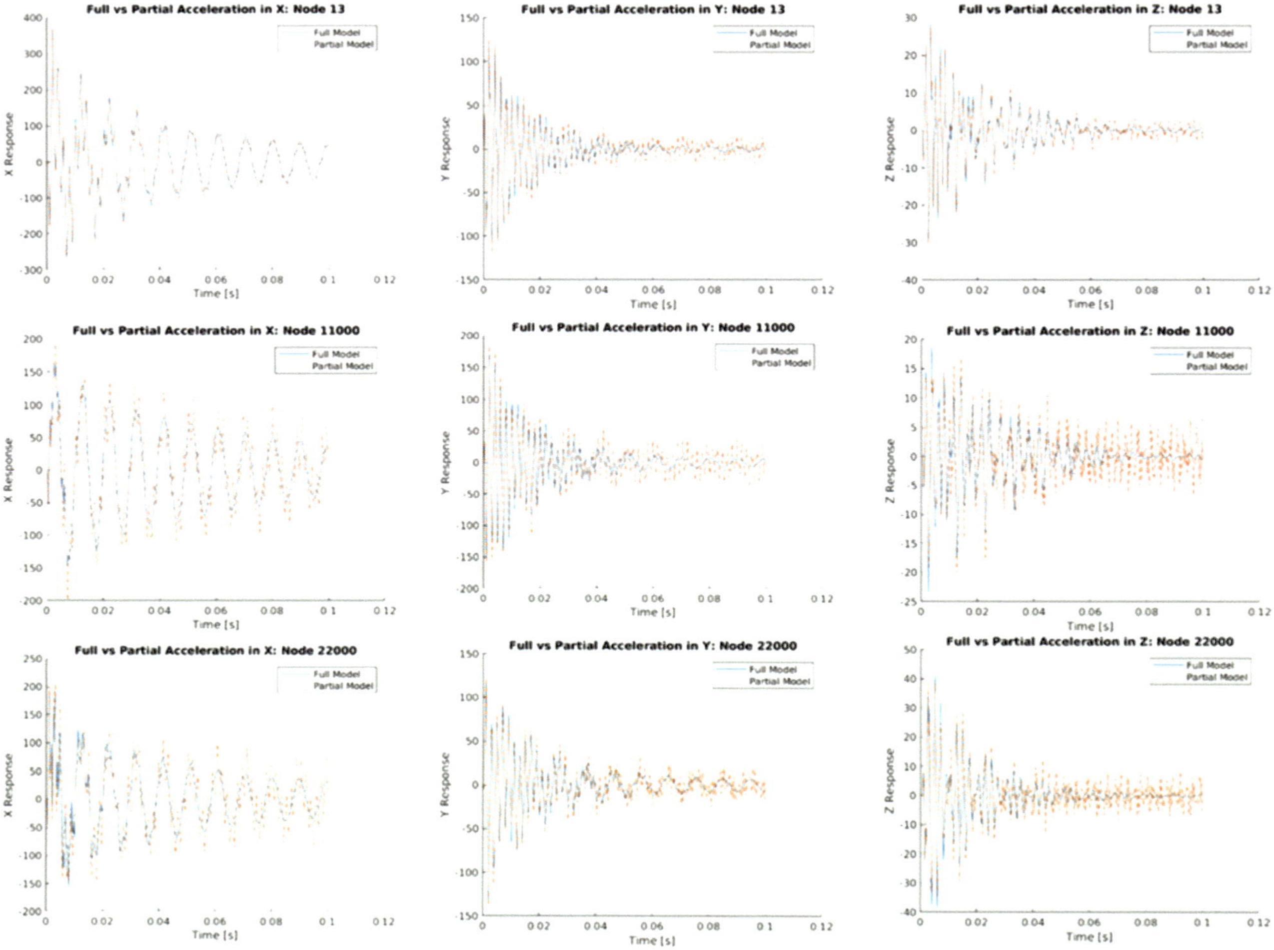

Fig. 7 Reapplied force response comparison at top corner of the RC

of a component is unknown, which is often the case in practice. For example, the response of a specific component in a flight vehicle might be desired to qualify the component. However, force reconstruction will not help predict the response of that component if the dynamics of the connection to the flight vehicle is not also quantified. Error in the defined boundary conditions of a component will also propagate to the component's response.

Conclusion

This work investigated how errors in applied loads to FEA models propagates to the predicted responses. First, the BARC model was used to quantify how errors in applied loads propagate to the predicted response using a common force estimation technique. The exercise showed that the error in response can be significantly larger than the error in the input, and that even reasonable estimates of input force can produce huge response errors. Next, the same BARC model was used to illustrate how uncertainties in the boundary conditions of a component of interest will also propagate to errors in the predicted response, even if the force reconstruction is perfect. Future work should focus on developing techniques and promoting best practices for analysts to characterize errors in both applied forces and boundary conditions.

References

1. "FEMtools Model Updating: An Integrated Solution for Structural Dynamics Simulation, Finite Element Model Verification, Validation and Updating" FEMtools. https://www.femtools.com/products/ftmu.htm.

2. Akcelik, Volkan, Aquino, Wilkins, Kurzawski, Andrew, McCormick, Cam, Sanders, Clay, Smith, Chandler, Treweek, Ben, and Walsh, Timothy. Inverse Methods - Users Manual 5.8. United States: N. p., 2022. Web. doi:10.2172/1874427. "Fusion of Simulation, Experiment, and Data (FuSED) User's Manual – 5.20.". United States.
3. Fowler, D., Parker, S., Cleal, M., Avitabile, P., Logan, P., Roettgen, D. R., ... & Kuether, R. J. (2019). Force Reconstruction at Mechanical Interfaces (No. SAND2019-11669C). Sandia National Lab.(SNL-NM), Albuquerque, NM (United States); Sandia National Laboratories
4. Chen, S., Niezrecki, C., Avitabile, P., Wu, N., & Wang, X. (2016, June). Numerical simulation and dual experimental mapping of acoustic field generated by ultrasonic transducers. In INTER-NOISE and NOISE-CON Congress and Conference Proceedings (Vol. 252, No. 2, pp. 849-856). Institute of Noise Control Engineering.
5. Crane, Nathan, Day, David, Dohrmann, Clark, Stevens, Brian, Lindsay, Payton, Plews, Julia, Vo, Johnathan, Bunting, Gregory, Walsh, Timothy, and Joshi, Sidharth. 2022. "Sierra/SD - User's Manual - 5.10.". United States. https://doi.org/10.2172/1887938.https://www.osti.gov/servlets/purl/1887938.

Chapter 4
Toward Generalized MIMO Random Vibration Specifications

Ryan Schultz and Garrett Nelson

Abstract For decades, single-axis vibration specifications have been derived using generalized approaches that may misrepresent the exact test article dynamics and detailed motion that would be experienced in a particular environment. These approaches were developed to represent a conservative envelope of responses which accounts for multiple sources of variability including unit to unit differences and variations in the service environment loading. Testing to generalized specifications has many benefits, including broad applicability to different test units and the ability to encompass multiple environments with one specification. However, this generalization comes at the cost of true replication of field environment responses in the laboratory. Rather than perfect field replication, the intent of most tests is to subject the test article to vibration that is similar to expected field environments and scaled to a conservatively high level in order to sufficiently stress the item.

Despite the growing popularity of multiple-input/multiple-output (MIMO) vibration testing, most MIMO tests utilize replication-style specifications that lack the ability to account for uncertainties often desired for acceptance, qualification, or production testing as is done for their single-axis counterparts. This is likely for two reasons. First, because MIMO testing can accurately replicate complicated environments, accuracy is made a primary goal of the test. Second, there simply are not many available methods to create or utilize generalized specifications for MIMO tests. This paper presents a method to create a generalized MIMO random vibration specification by incorporating an ensemble of responses and statistical conservatism in much the same way as traditional, single-axis specifications. However, the resulting specification should not be directly imported into a MIMO controller for use. A second step is presented which maps the generalized specification to inputs to the MIMO controller, capturing all the detailed, narrowband, test article-specific information required for a good laboratory test that still accounts for additional sources of variability.

Keywords MIMO · Random vibration · Vibration testing · Test specification · Multi-axis

Introduction

In MIMO vibration testing, a control system determines a set of shaker drives that will minimize response error at several response locations on the device under test (DUT). With a good test design and setup, MIMO control can accurately match the acceleration response at various locations or degrees of freedom (DOFs), potentially exactly mimicking the complicated response patterns that the DUT may see in a service environment. While a perfectly accurate test is appealing, it is often difficult or impractical to achieve for several reasons:

- There may be too few shakers (inputs) or shakers may be in poor locations
- There may be too few accelerometers (outputs) or accelerometers may be in poor locations
- The test specification may be incompatible with the DUT dynamics
- There may not be enough relevant field data available to make a MIMO-compatible specification

The first two issues can often be remedied with good test design methods. The last two issues are the real problems that make MIMO testing difficult to apply to many systems. Specifications must be compatible with DUT dynamics for the test to be controllable and for the test to be run efficiently. For example, if the specification has a peak response at 100 Hz, but the DUT has modes at 50 and 200 Hz, then it will likely be difficult to drive a peak at 100 Hz while suppressing motion at 50 and 200 Hz. Each difference in spectral shape can require significant drive energy and makes the test difficult to run, especially at high test levels, as the shakers need to force the DUT to move in ways that it does not naturally want to.

Ryan Schultz · Garrett Nelson
Sandia National Laboratories, P.O. Box 5800 – MS0346, Albuquerque, NM, 87185
e-mail: rschult@sandia.gov; gdnelso@sandia.gov

Alexandra Karlicek et al. (eds.), *Dynamic Environments Testing, Vol. 7*, Conference Proceedings of the Society for Experimental Mechanics Series, https://doi.org/10.13052/9788743801528-4

Having enough field data to support MIMO specifications is another challenge. For the field data to be useful in traditional MIMO test specification derivation, the field data needs to be derived from a very similar system (similar frequencies, mode shapes, and damping as the DUT), have similar load paths as the MIMO test (shaker locations), and have similar boundary conditions. There are very few field tests that were designed and performed with MIMO testing in mind, and as a result, most of the available field test data will not be very useful for deriving MIMO specifications.

Another gap with traditional MIMO specification derivation methods is that they rely on a single data set (one test system, one environment). While this may allow the response of that particular test to be perfectly replicated in the laboratory, that is not always the goal of vibration testing. Often, the tests are not replication tests (where the field response is matched perfectly) but are qualification tests, where the goal is to subject the DUT to a conservatively high vibration environment that encompasses or represents a set or ensemble of environments, test articles, or test conditions. Statistics are computed on an ensemble of field data to estimate the maximum predicted environment (MPE) [1, 2], which is then used in test specifications to ensure the DUT can survive "worst case" conditions expected during a field environment.

What is needed for MIMO vibration specifications are techniques that can maximize the usefulness of the available field data and utilize multiple data sets together to estimate statistics. These techniques should:

- Allow multiple data sets to be combined, even if the gauge locations differ from test-to-test (e.g., allow for spatial or zone averaging) [1].
- Be robust to changes in system dynamics, allowing for unit-to-unit or test-to-test shifts in modes without overly penalizing levels or variance estimates used in MPEs (e.g., allow for frequency averaging).
- Result in specifications that are testable by typical MIMO control systems (e.g., can be translated or mapped to narrowband power spectral densities (PSDs)).
- Allow for a small number of specifications (individual PSDs) to meet the intent of the test. A small number of frequency-averaged PSDs is easy to communicate, document, and compare against test results.

The benefits of simplification and general applicability are highlighted in Figure 1 and Figure 2 which show traditional, replication-style MIMO specifications (Figure 1) vs. the proposed, generalized MIMO specifications (Figure 2). In traditional MIMO specifications, there is a unique narrowband PSD for each control DOF. With tests often using 20 or more control DOFs, the result is many, many specification PSDs. In contrast, with a generalized specification framework, specifications can be defined at zones and there can be far fewer zones than control DOFs. Those specifications can be defined as frequency-averaged quantities, further reducing the number of specification PSDs and increasing the applicability of those specifications to other variations of the system. It may even be possible to reduce the generalized MIMO specification PSDs for a system down to a familiar breakpoint table which can be reported in test documents as the size of the specification goes down to: N_{zones} x $M_{frequency\ bands}$ where $N_{zones} << N_{control\ DOFs}$ and $M_{frequency\ bands} << M_{frequency\ lines}$. Take for example a case with 20 control DOFs, 2000 frequency lines, 5 zones and 10 frequency bands. The traditional MIMO specification

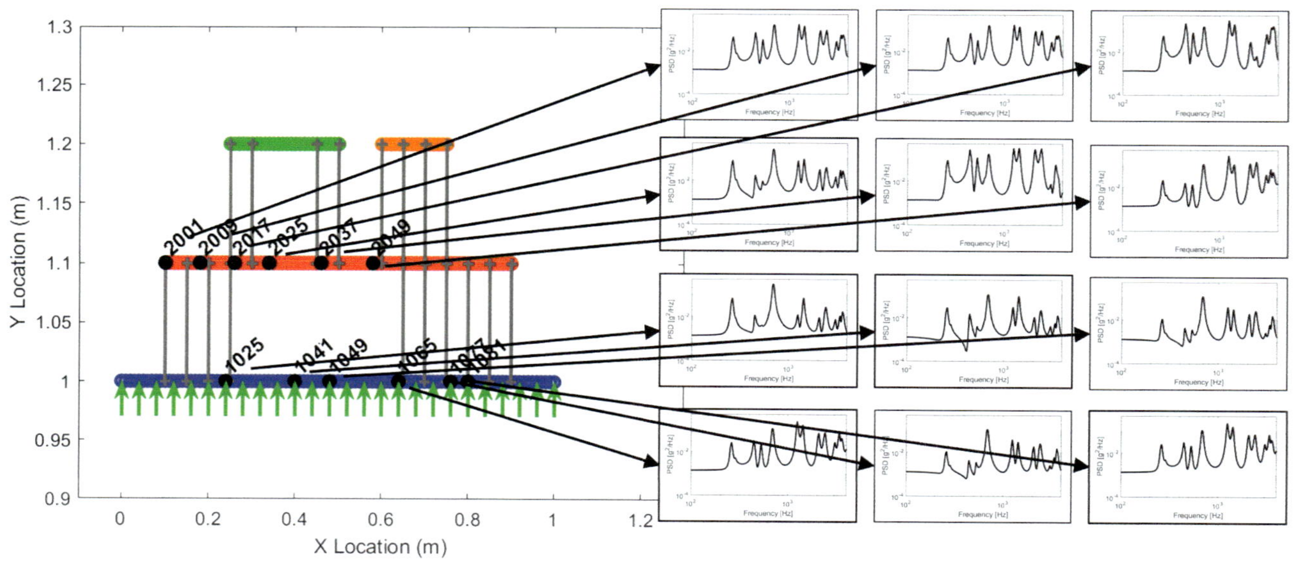

Fig. 1 Traditional MIMO specification example where each DOF has a unique specification PSD.

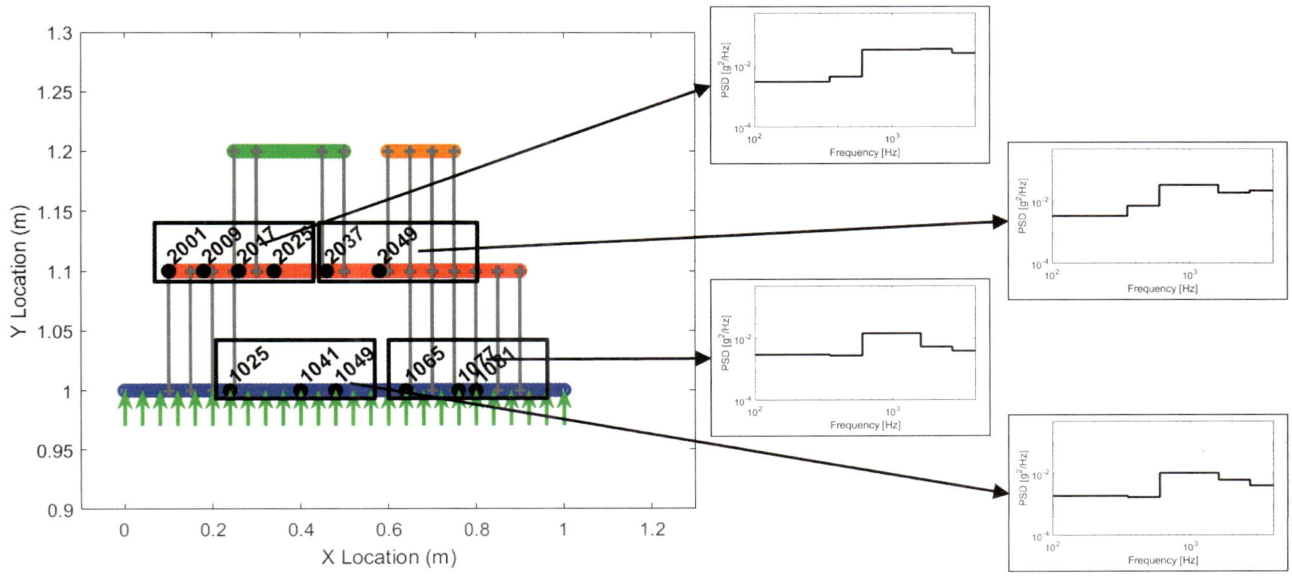

Fig. 2 Generalized MIMO specification example where DOFs are averaged into zones and each zone has a frequency-averaged specification PSD.

would entail 40,000 unique PSD values, ignoring cross-terms, (20 x 2000) whereas the generalized MIMO specification would only require 50 unique PSD values (5 x 10).

The overall goal in this work is generalized MIMO specifications: if MIMO specifications can be generated with those features, then they can likely be more broadly applied to more test articles, encompass more environments, and make use of existing field data sets. This paper presents an approach to enable generalized MIMO random vibration specifications by mapping from the generalized specifications to the specific PSDs required as control inputs to the MIMO controller. First the approach is discussed, including some gaps and assumptions. Next, an example problem is shown to demonstrate how this can be useful. Finally, results and conclusions are presented along with recommendations for future work.

Proposed Method For Generalized MIMO Specifications

There are two phases of the problem to consider: first, creating generalized MIMO specifications and then second, being able to use those generalized MIMO specifications in a test. The intent is to create a set of generalized specifications which may be derived from a various versions of a fielded system and various sets of field vibration data, and then to be able to use that for a specific DUT, which may have different dynamics. A translation or mapping is required to go from the generalized specification (which may be a small set octave-averaged, zone-averaged PSDs) to a set of MIMO controller PSDs (which may be a larger set of narrowband PSDs which reflect DUT-specific dynamics). Note that the generalization methods described in section 2.1 should not be used without the mapping methods described in section 2.2, they are intended to be used together to create the most representative test conditions.

Creating Generalized MIMO Specifications

The first step in the proposed method is to create a generalized MIMO PSD by averaging over gauges, frequencies, and data sets. Averaging over gauges allows the number, location, and direction of gauges to vary somewhat test to test while still being reasonable to compile together. This softens the problem by saying the average response of a set of gauges in a region or zone of the DUT should match an averaged specification. The same zone-averaged specifications could be applicable to multiple variants of a DUT even if the gauge layouts change somewhat test to test. It is useful to note that many single-axis specifications are exactly this, where there is just one zone (the base/input to the system), and multiple sensors (all mounting locations) are averaged together to represent this zone.

Averaging over frequencies allows the dynamics to vary between field test to field test and field test to lab test without unnecessarily forcing the dynamics of the field system onto the lab DUT. The intent here is to allow the DUT to vibrate as it naturally would in service (and not as some other field-test system vibrated in service). Frequency-averaging can be accomplished by using octave averaging or any band-wise frequency averaging scheme. Note that the term octave averaging is used in this paper, even if the frequency bands are not necessarily octave spaced. Ideally, the frequency bands would encompass dynamic features of interest and allow those features to change in frequency somewhat without moving to an adjacent band. This softens the problem by saying the average response across a range of frequencies in a given octal band should match rather than saying every single frequency line should match field vs. lab. Note that there will be a balance of generalization vs. accuracy in choosing frequency bands. Too narrow of a band means that features can cross band edges and are no longer mappable from field to lab. Too wide of a band means that too many features will be in the band and may average together and make it difficult to enforce desired motion on the lab DUT.

Specifications can be made conservative by computing statistics (mean and variance) over multiple data sets and using those statistics to estimate the MPE. It is important to note that the statistics may be very different if computed with narrowband PSDs vs. octave-averaged PSDs (i.e., the variance in octave-averaged PSDs is much lower than narrowband PSDs due to frequency shifting). As a result, for MIMO applications it is recommended for statistics to be computed on octave-averaged quantities.

An example process for creating generalized MIMO random vibration specifications would be:

1. Compile the set of relevant field test data. Any test on a similar system to the DUT in a similar environment should be included to maximize the use of available field test data.
2. Determine frequency bands for octave averaging. These bands should encompass dynamic features in the field systems and field data and the lab DUT. Ideally, bands would have similar features (i.e., modes) for both the field and lab systems. Bands can be determined by examining field test PSDs and lab test system identification or frequency response function (FRF) data.
3. Determine groups of gauges to put into zones for spatial averaging. These should be groups with similar locations on similar components or hardware features, but the gauge locations within each zone can be different from field test-to-field test or field test-to-lab test. The goal is to have the response be similar across gauges in the zone. Appropriate zones can be determined by inspection (knowledge of the system construction) or by examining modal or field test data.
4. Apply octave averaging to the field test data, resulting in, for example, $1/3^{rd}$ octave PSDs for each field test at each DOF.
5. Apply zone averaging to the field test data, resulting in, for example, a $1/3^{rd}$ octave PSD for each zone.
6. Compute the mean and variance from the set of zone-averaged, frequency-averaged PSDs across all field test data sets. Do typical due diligence in understanding the distribution, identifying outliers, etc.
7. Use the mean and variance estimates to estimate the MPE for a given level and confidence interval, for example P95/50. Add additional scaling to account for deterministic factors such as known system-to-system differences. The resulting scaled P95/50 level averaged PSDs would be the MIMO specifications.

This process is illustrated in Figure 3 and Figure 4 which shows two variations of a system, System A and System B, and the associated field data. Suppose field data only exists for System A but a lab MIMO test needs to be performed on System B. System B has somewhat different dynamics and instrumentation from System A. Those differences are highlighted in Figure 5a where the response of each system is overlaid and shifts in dynamics in the ensemble of data are distinctly visible. Simply controlling the lab test of System B to responses from System A would result in an inaccurate and inappropriate test – forcing System B to response exactly as System A had in a field environment and not the way System B would have if it had been in the same environment. Instead, the System A field data can be averaged over zones of like-DOFs (shown in the black boxes in Figure 3) and averaged over frequency bands (shown in the averaged PSD in Figure 5b). Those averaged quantities are then transferrable to the lab test of System B.

Converting The Generalized MIMO Specifications To Testable PSDs

The averaged or generalized MIMO specifications are not directly usable, meaning you cannot simply import them into a controller and run a meaningful test. This is for a few reasons. First, controllers need narrowband PSDs as specification inputs, so the octave-averaged specification needs to be converted to narrowband. Second, the specification is in terms of zone-averaged PSDs, so it needs to be converted to the DOFs in the DUT. And third, we want to incorporate the dynamics

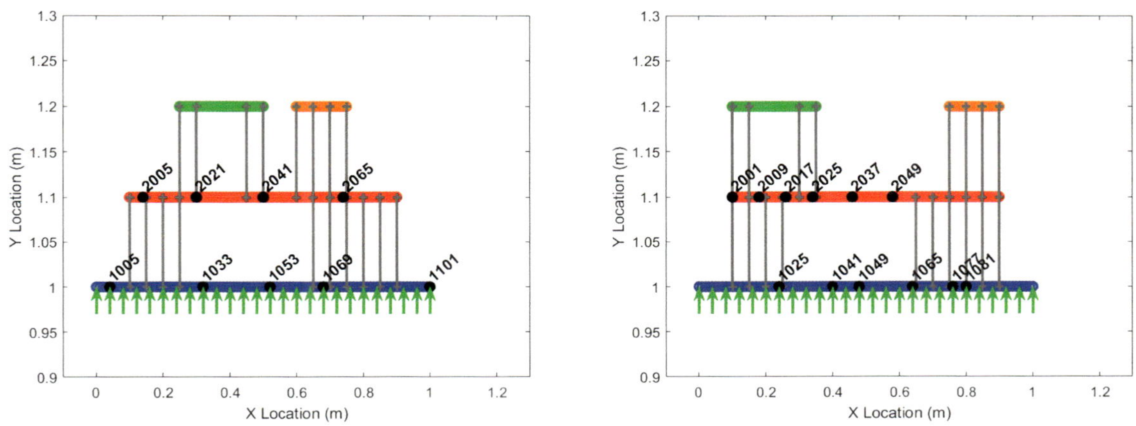

(a) (b)

Fig. 3 Two configurations of an example multi-component dynamic system with DOF grouped into zones as shown in black boxes. (a): System A. (b): System B.

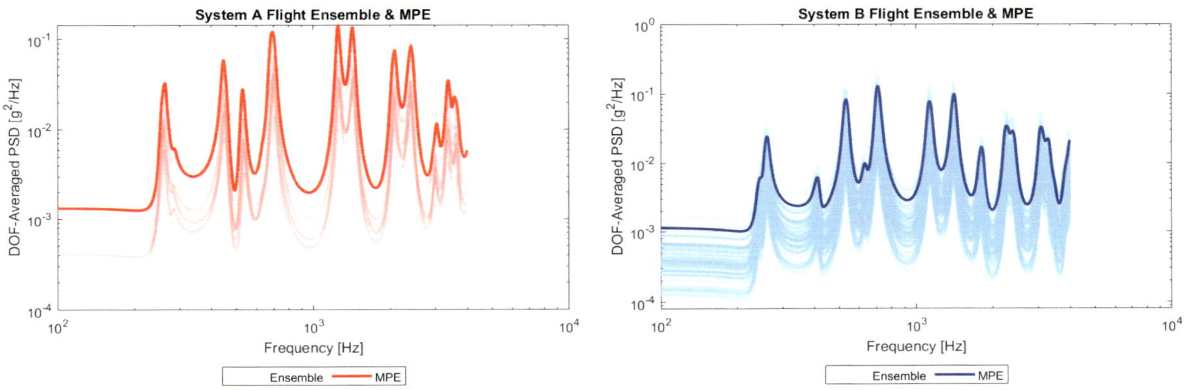

(a) (b)

Fig. 4 Example field response and MPE for System A and B. (a) Ensemble of field data and a P95/50 MPE for System A. (b) Ensemble of field data and a P95/50 MPE for System B.

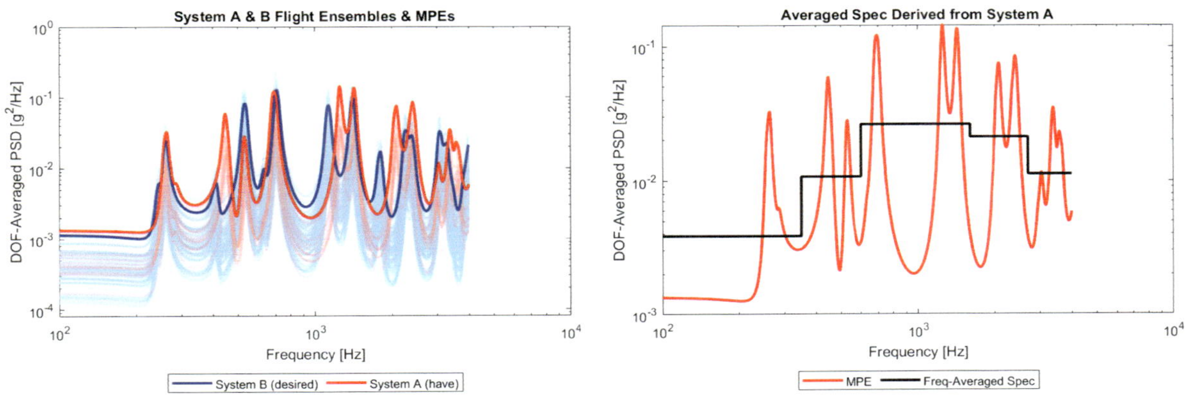

(a) (b)

Fig. 5 Example field response, MPE, and specification for System A and B. (a) Comparison of System A and B field data to show dynamic differences. (b) Example System A MPE and octave-averaged PSD.

of the DUT as much as possible, so it is helpful to use some DUT test response as a basis for converting from the specification octave PSDs to the narrowband PSDs needed for the controller. If available, DUT field response would be the most representative, but DUT lab system identification response can work in many cases as well.

Generalized specifications are beneficial in that they are simple, averaged representations of response. However, that simplification requires we "fill in the blanks" with lab DUT-specific information to perform a meaningful test. This can be done by utilizing the DUT system identification test data (PSDs) as a set of basis functions. Those DUT PSDs contain all the DUT-specific frequency, mode shape, and damping information we want to incorporate into the MIMO test.

The basis PSDs allow the generalized PSDs to be mapped to narrowband PSDs with the following process:

1. Create the set of basis PSDs for the lab DUT (a set of $N_{control\ DOFs}$ x $M_{frequency\ lines}$ narrowband PSDs). This most easily obtained from system identification test data on the lab DUT, but may come from other sources.
2. Octave average the basis PSDs to the same frequency bands as the generalized MIMO specification PSDs.
3. Zone-average the octave-averaged basis PSDs, making the averaged basis PSDs comparable to the generalized MIMO specification PSDs (size N_{zones} x $M_{frequency\ bands}$)
4. Compute the difference between the generalized MIMO specification PSDs and average basis PSDs as a set of scale factors (size N_{zones} x $M_{frequency\ bands}$).
5. Interpolate the scale factors to narrowband frequency lines (size N_{zones} x $M_{frequency\ lines}$). This can be done with, for example, spline interpolation on each zone's PSD scale factors.
6. Apply the narrowband scale factors to the basis PSDs for each DOF in each zone where the DOFs in the same zone are scaled the same. This results in a set of basis PSDs which are scaled to match the generalized specification in terms of location (zone) and frequency band (size $N_{control\ DOFs}$ x $M_{frequency\ lines}$)

Note that the objective of the above process is to scale the narrowband basis PSDs such that they match the zone-averaged and octave-averaged PSDs in the specification. The interpolation of scale factors from octave to narrowband can cause the octave-averaged scaled basis PSDs to be slightly different from the desired specification PSDs, so the process can be repeated for a few iterations and get better agreement.

The mapping from generalized averaged specification PSDs to testable narrowband PSDs is shown in Figure 6, Figure 7, and Figure 8. Figure 6 shows the averaged specification PSD along with the narrowband PSD it was derived from (the System A MPE PSD) in comparison with the System B basis PSD. Note the differences in spectral content between the System A MPE and System B basis PSDs. Also note how the frequency bands capture similar features in both sets. For example, the band around 1000 Hz captures three peaks in both the System A MPE and the System B basis.

Figure 7 shows an example of the octave and narrowband scale factors used to scale the basis PSD to the averaged specification PSD, including each scaling iteration. Figure 8 shows the resulting mapped specification in both narrowband and octave-averaged form compared with the basis PSD and desired octave-averaged specification. Two key observations:

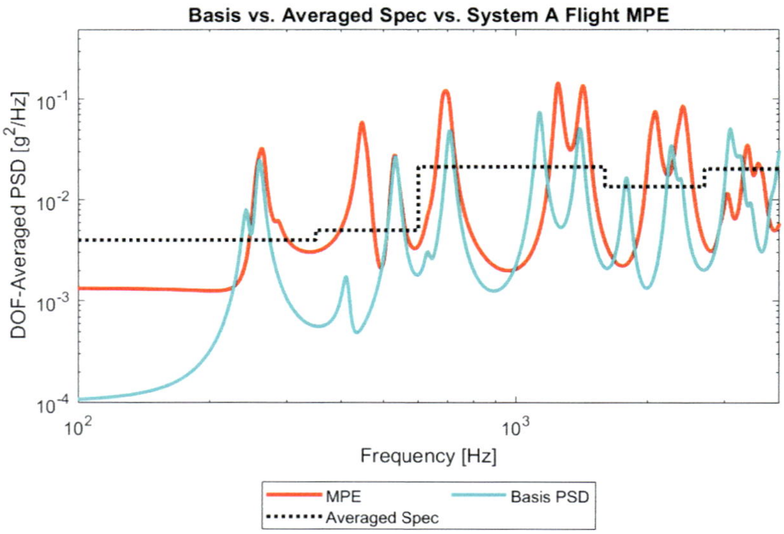

Fig. 6 Comparison of the System A MPE narrowband PSD vs. the System B basis PSD with the octave specification PSD to show band edges .

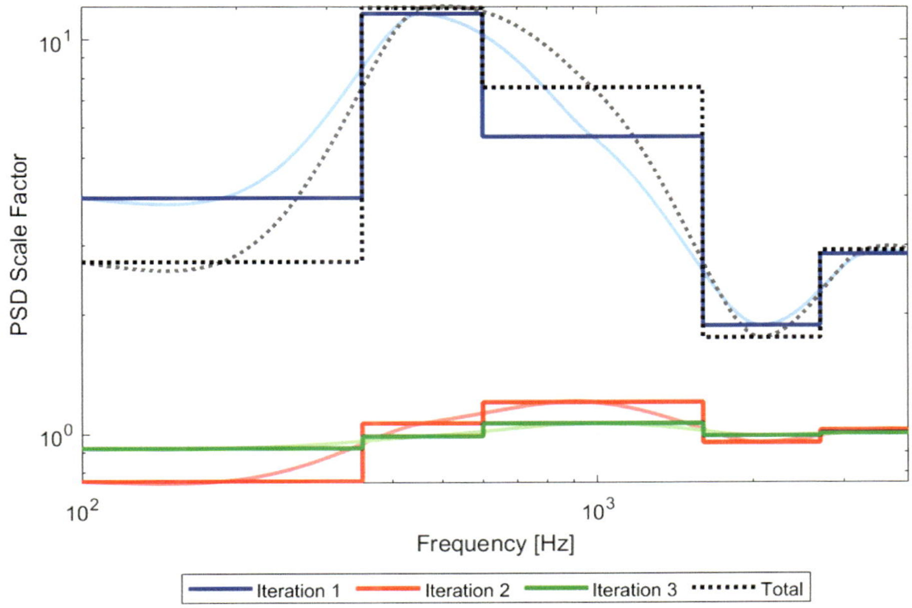

Fig. 7 Scale factors between octave specification and octave-averaged basis and those scale factors interpolated to narrowband frequency lines. Straight lines indicate octave band scale factors and smoothed curves indicate the narrowband scale factors from interpolating the octave scale factors. Black is the total scaling after all iterations. Colors show individual iterations which converge to unity scaling after a few iterations.

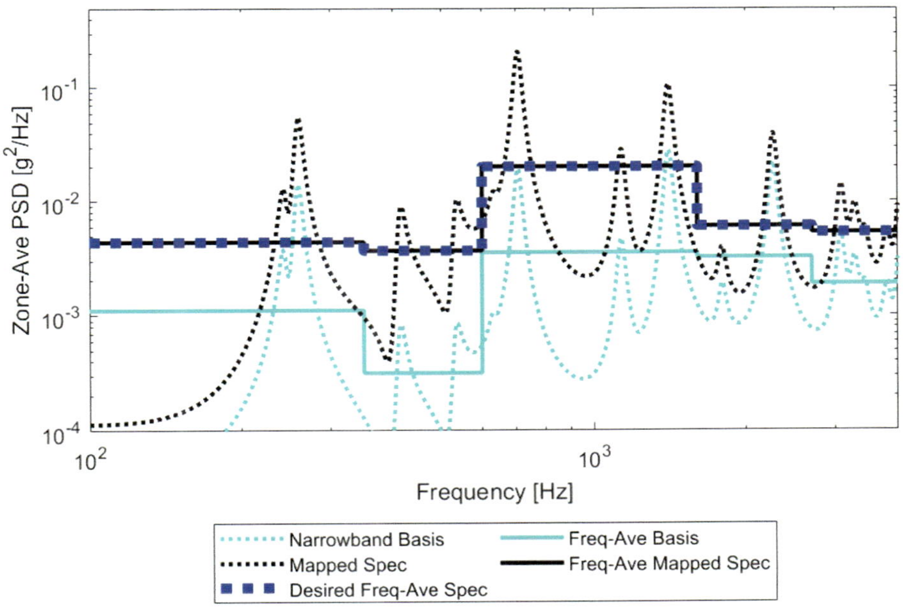

Fig. 8 Basis PSD scaled to match the averaged specification PSD.

first, the mapped specification matches the desired specification at each octave band, meaning the energy is appropriately captured frequency band by frequency band. Second, the details in the System B basis PSD are generally preserved after scaling, indicating that the System B modes will be appropriately excited during MIMO testing. Of course, there is the caveat that the basis PSDs must reasonably match the field PSDs for those modes to be appropriately excited, so some attention should be paid to obtaining useful basis PSDs. Importantly, these mapped specifications will not force System B to vibrate exactly like System A did in flight, rather they will force System B to vibrate approximately like System B would have in flight.

Simulation Demonstration

The generalized MIMO specification and mapping process will now be demonstrated on an example numerical system. Two versions of the system model were created, System A and System B, as shown in Figure 3. They have nominally the same construction with four beams connected via spring elements. The bottom beam has a pressure load applied in the field environment (e.g., the outer surface of a missile in flight). System B has a different configuration of the green and orange components compared with System A, as well as some minor changes to stiffness and damping. The intent is for System A and B to be similar in overall character while having some minor differences in modes that may be expected in real world examples, reflecting the differences due to unit-to-unit variability or different design iterations. A field environment is simulated with a partially-correlated pressure load applied to the bottom beam [3, 4], to simulate a free-flight environment. Multiple field tests are simulated with randomly varying load amplitude to provide an ensemble of responses. The MPE is then computed from the ensembles as shown in Figure 4. Note that in practice, there will not be field data for System B (the DUT), but here it is useful to have that truth data for comparison with the mapped specifications and test results.

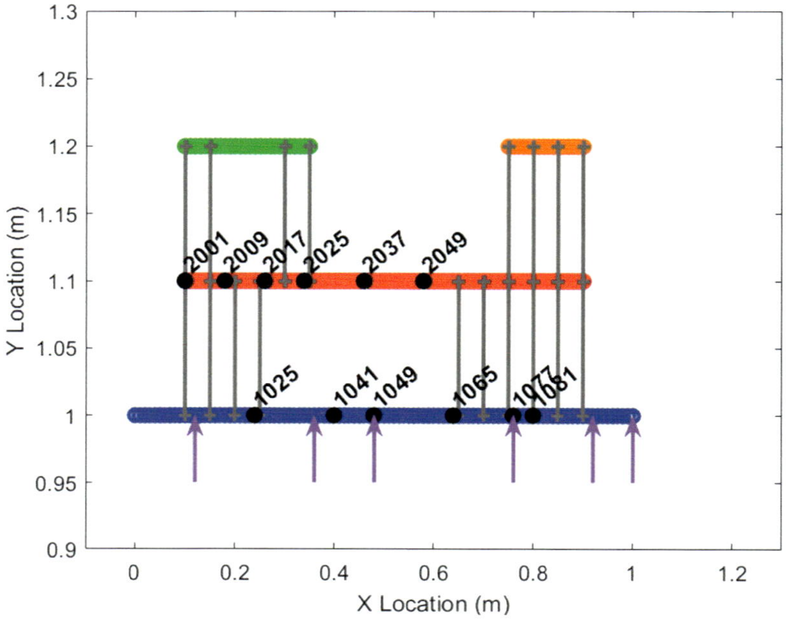

Fig. 9 Six shakers located on System B.

A MIMO test on System B is simulated by applying six point forces to the bottom beam in the same way that shakers would be set up in an Impedance-Matched Multi-Axis Test (IMMAT) [5, 6, 7]. The number and location of forces (shakers) is important but outside the scope of this paper. In this case, six forces at the locations shown in Figure 9 were sufficient to accurately mimic the response of the simulated field environment.

The basis PSD is obtained by simulating a system identification test on System B, where uncorrelated inputs were applied at the six shaker locations. Next, the basis PSDs are averaged over frequencies and over DOFs per the zones shown in Figure 3, where the field test has 9 DOFs and the lab test has 12 DOFs. Next, the averaged specification, derived from the System A field tests, is mapped to System B test control PSDs by scaling the System B basis PSDs. An example of this scaling is shown in Figure 8. The result of that mapping is a set of 12 narrowband PSDs which comprise the System B MIMO test control PSDs (PSDs that would be provided as specifications to the controller).

The System B MIMO test is simulated using the System B FRFs (from the simulated system identification test) and the mapped test control PSDs. The overall results of that test are shown in Figure 10, where the narrowband and frequency-averaged specification is compared with the test response. The test controlled reasonably close to the specification PSD over all DOFs and zones of DOFs, though perfect control is not always possible due to test setup details (such as shaker locations) or incompatibility of the specifications with the DUT dynamics (such as due to disparate scaling DOF to DOF or zone to zone).

Benefits of this generalized specification and mapping approach are shown in Figure 11 and Figure 12 where the test results are compared with the true System B field response ensemble and MPE. The traditional MIMO specification approach is shown in Figure 11, where the test control PSDs (specifications) are derived directly from the System A flight MPE. As a result, the test response closely mimics System A responses and dynamics, which does not match the true System B flight response. Several peaks are forced to respond at different amplitudes and different frequencies. In contrast, when specification PSDs are generalized and mapped via the System B basis PSDs to get the control PSDs, as shown in Figure 12, the resulting test response is a much closer match to the true System B flight response. Peak frequencies and other spectral features are maintained. Of course, there will be some amplitude error due to the mismatch in response amplitude between System A and B and error in the mapping process, though it should be noted that the amount of error will be problem dependent.

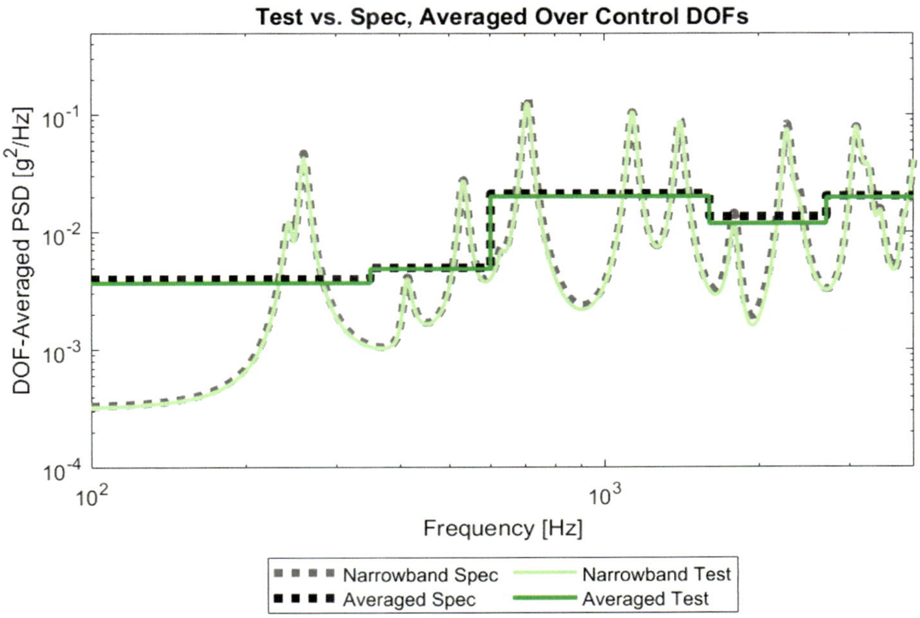

Fig. 10 Simulated MIMO test results shown as the response PSD averaged over all 12 control DOFs.

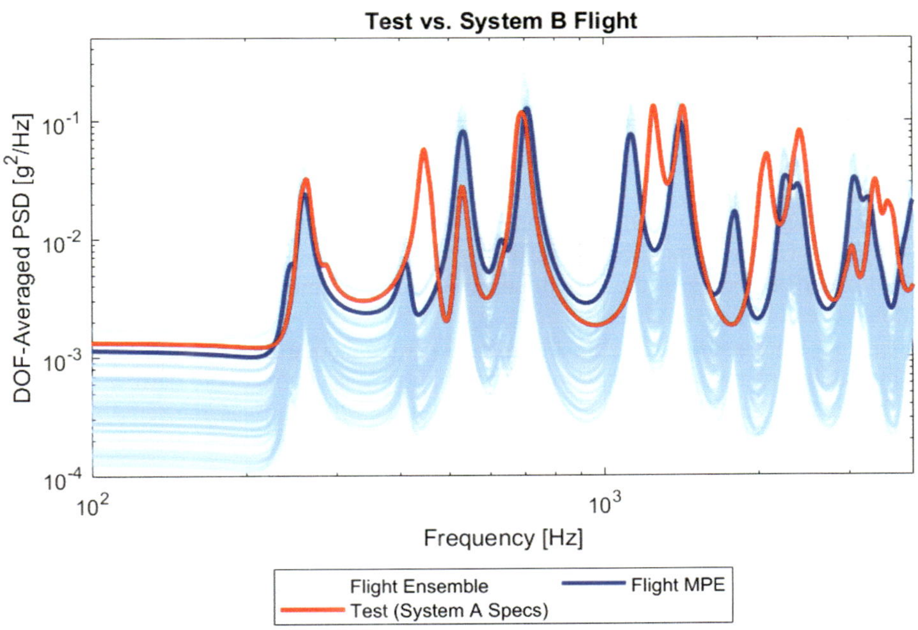

Fig. 11 Simulated MIMO test results compared to the System B field test ensemble and MPE with MIMO specification PSDs taken directly from System A field test data.

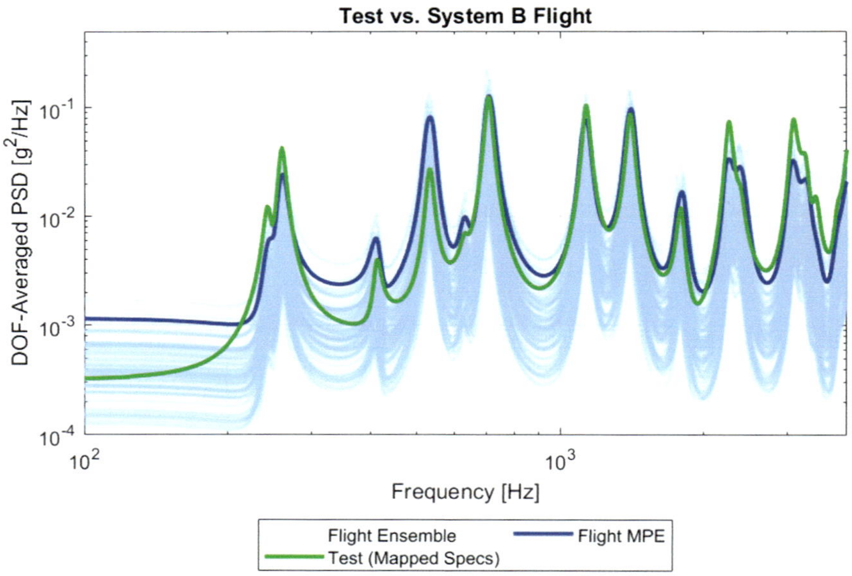

Fig. 12 Simulated MIMO test results compared to the System B field test ensemble and MPE where the generalized MIMO specifications process is used.

Discussion

The presented change to how MIMO specifications can be defined opens up several opportunities for the test engineer to change settings or methods that affect test results. In general, it is better to have more options with MIMO as it gives the test a better chance to run efficiently and accurately compared with a traditional approach where much of the design and nuance is buried in the original field data. That said, several things should be considered before using this process in a test.

The generalized PSDs described in this paper are only auto-power spectral densities and cross-terms are ignored as the concepts of frequency and zone averaging are likely not applicable to coherence and phase. Cross-terms must be incorporated with the control PSDs in order for a MIMO test to be run. These could come from the basis PSD data, or from an example field test, or from some other method [5, 8, 9]. In any case, the test engineer should understand what cross-term approach is appropriate and how it may affect the DUT response.

The choice of the basis PSDs will have a significant impact on the test and every effort should be made to make the basis PSDs as similar as possible to the expected DUT field response. This includes DUT dynamics, boundary conditions, load paths, input correlations, and levels as these factors will all affect DUT response. Naturally, some of these are easier to control or implement than others, but all should be in mind when designing a test. While the example here used uncorrelated inputs to create the basis PSDs, that may not be the right choice for some systems or environments.

Care should be taken when defining DOF zones and frequency bands to ensure similar features are grouped together. If defined poorly, responses could be mapped to the wrong locations, or the wrong spectral features could be amplified or attenuated. A detailed understanding of the system construction, instrumentation configuration, and system dynamics is required to ensure this process works as intended.

The process of mapping from averaged specifications to narrowband controller PSDs is a bit complicated and introduces some assumptions, including the method for interpolating from octave scale factors to narrowband scale factors. That said, it is believed this is better than simply assigning all frequency lines within a band the same PSD value as that will force an unnatural response on the DUT [10]. The test engineer should understand the various assumptions and make appropriate tradeoffs to ensure the intent of the test is met.

Conclusions

A limitation to widespread use of MIMO vibration testing is a lack of specifications to run tests, mostly due to the various requirements needed in a good MIMO test specification. This work aims to soften those requirements while still meeting the intent of the test: to subject the DUT to field-relevant vibration environments while letting the DUT dynamics exist as they would in service. This was accomplished by considering the MIMO specification to be an averaged quantity over both spatial locations and frequency ranges. In this way, DUTs with variations in instrumentation configuration or dynamics can still utilize the same test specifications. This has a few important benefits: it reduces the complexity of the specifications, allowing them to be tabulated or easily plotted, it does not require bespoke specifications for each test or DUT, and it allows multiple disparate field data sets to be combined into one set of specifications, maximizing the usefulness of limited field data.

These generalized specifications were then made usable via a novel mapping process where a set of DUT-specific basis PSDs are used to interpolate the generalized specifications across DOFs and frequency lines, resulting in a set of narrowband PSDs that can be imported as control PSDs in any MIMO controller. Importantly, as this mapping process utilizes DUT-specific basis PSDs, the control PSDs will still maintain the DUT's frequencies and mode shapes, allowing the DUT to vibrate naturally, as it would in service.

Of course, this step toward generalized MIMO specifications comes with drawbacks. Primarily, more effort is required to design the test setup to mimic in-service conditions as closely as possible. As the test setup will affect the basis PSDs, and the basis PSDs affect the control PSDs and therefore the test results, it is imperative that the test setup is designed with this process in mind. Pre-test predictions and sensitivity studies should be performed to verify the configuration and understand design tradeoffs. Similarly, by softening the constraints at the test specification, there is more user interaction and many options to change in the specification mapping process, which necessitates an expert in the loop.

Acknowledgments Sandia National Laboratories is a multimission laboratory managed and operated by National Technology and Engineering Solutions of Sandia, LLC, a wholly owned subsidiary of Honeywell International, Inc., for the U.S. Department of Energy's National Nuclear Security Administration under contract DE-NA0003525.

References

1. National Aeronautics and Space Administration, 11Dynamic Environmental Criteria. NASA Technical Handbook NASA-HDBK-7005," 2001.W409W9360
2. "Test Requirements for Launch, Upper-stage and Space Vehicles, SMC Standard SMC-S-016," Air Force Space Command, Space and Missile Systems Center Standard, 2014.W409W9360
3. F. Jacobsen, "The diffuse sound field," Acoustics Laboratory at the Technical University of Denmark, 1979.W409W9360
4. M. Kuster, "Spatial correlation and coherence in reverberant acoustic fields: Extension to microphones with arbitrary first-order directivity," The Journal of the Acoustical Society of America, vol. 123, no. 1, 2008. W409W9360
5. P. M. Daborn, "Scaling up of the Impedance-Matched Multi-Axis Test (IMMAT) Technique," in Proceedings of IMAC XXXV, the 35th International Modal Analysis Conference, Garden Grove, CA, 2017. W409W9360
6. C. Roberts and D. J. Ewins, "Multi-axis vibration testing of an aerodynamically excited structure," Journal of Vibration and Control, vol. 24, no. 2, pp. 427–437, 2018. W409W9360
7. P. M. Daborn, C. Roberts, D. J. Ewins and P. R. Ind, "Next-generation random vibration tests," in Proceedings of IMAC XXXII, the 32nd International Modal Analysis Conference, Orlando, FL, 2014. W409W9360
8. U. Musella, G. D'Elia, A. Carrella, B. Peeters, E. Mucchi, F. Marulo and P. Guillaume, "A minimum drives automatic target definition procedure for multi-axis random control testing," Mechanical Systems and Signal Processing, vol. 107, pp. 452–468, 2018. W409W9360
9. D. O. Smallwood, "A revised algorithm for minimum input trace to a multiple-input/multiple-output system (MIMO) to match the output autospectral densities," in Proceedings of the 85th Shock and VIbration Symposium, Reston, VA, 2014. W409W9360
10. R. Schultz and G. Nelson, "Techniques for modifying MIMO random vibration specifications," in Proceedings of IMAC-XL, the 40th International Modal Analysis Conference, Orlando, FL, 2022. W409W9360

Chapter 5
A Method for Response Replication at Component-Level in MIMO Random Testing

Enrico Proner and Emiliano Mucchi

Abstract This study focuses on Multiple-Input Multiple-Output (MIMO) vibration control tests, designed to accurately mimic a structure's operational vibration environment. In MIMO random testing, the Cross Spectral Density (CSD) between each pair of control channel plays a key role in the outcome of the test. In fact, the same Power Spectral Densities (PSDs) but with different CSDs may determine a different response on the Device Under Test (DUT) and different drives required to conduct the test. In this scenario, this paper proposes a methodology that aim to replicate a random vibration environment at the component level with minimal excitation. The approach pre-defines the phase and correlation of the reference Spectral Density Matrix (SDM) that maximize the response of the DUT. Then, the reference SDM is completed with the minimum PSD values ensuring that the structure's response in the laboratory aligns with the target vibration environment. This method enhances the consistency of laboratory tests with actual operational environments while minimizing the required energy, ultimately preserving the excitation system's integrity. The paper provides both a theoretical and an experimental analysis of the proposed methodology. The experiments are carried on specifically designed specimen excited by a multi-axis vibration environment, provided by a three-axis shaker.

Keywords MIMO · Random testing · Fatigue · Drive reduction · Environmental testing

Introduction

In the case of MIMO random control tests, the control target is a reference Spectral Density Matrix (SDM) comprising of Power Spectral Densities (PSDs), the diagonal elements, and Cross Spectral Densities (CSDs), out of diagonal elements [1]. The PSDs provide information about the energy of each control channels, the CSDs determine the correlation between different control channels, defined by means of phase and coherence [2]. Typically, only the PSD terms of the reference SDM are known. This is due to the legacy of single-axis testing procedures, which only require the PSD of the control channel, along side with the difficulty and impracticality of combining coherence and phase information from different operational conditions. In the recent past, several researches have focused on formalizing new methodologies to complete the reference SDM, exploiting the phase and coherence between the control channels to obtain more efficient tests [3, 4, 5, 6]. Moreover, the CSDs can have a significant effect on the DUT response [7, 8, 9]. At the authors knowledge, the available drive minimization techniques focus on minimizing the drives required to reach the test specifications, without considering the effects on the fatigue damage inflicted to DUT. Moreover, if the control location is on the DUT, the physical constraints of the DUT may not allow to achieve the minimum drives condition of the selected drive minimization technique. In this scenario, this paper proposes a novel MIMO random target generation technique, named *Minimum PSDs Method (MPM)*, which aims to minimize the energy required by the exciters to conduct the test while while preserving the damage inflicted to the DUT. The *MPM* predefines the phase and coherence between the control channels that maximize the response on the DUT for any given set of excitation PSDs. Then, the reference SDM is completed with the PSD values that generate the target vibration on the tested article. Because of the chosen phase and coherence the obtained PSDs are the minimum PSDs that generate the target vibration environment, dealing the same damage to component with minimal energy.

Enrico Proner · Emiliano Mucchi
Department of Mechanical Engineering, University of Ferrara, Italy, Via Saragat 1
e-mail: enrico.proner@unife.it; emiliano.mucchi@unife.it

Alexandra Karlicek et al. (eds.), *Dynamic Environments Testing, Vol. 7*, Conference Proceedings of the Society for Experimental Mechanics Series, https://doi.org/10.13052/9788743801528-5

The proposed Minimum PSDs Method

This section provides the mathematical implementation of a novel MIMO random target generation technique, namely *Minimum PSDs Method* (*MPM*). The concept of the method is to define the minimum PSDs of the reference SDM that allows to replicate a set of response PSDs on the DUT. The minimum PSDs can be obtained by using the Extreme Dynamic Response Method (EDRM) [7] to predefine the phase and coherence between the control channels. In fact, the EDRM aims to maximize the trace P of the response of the DUT, which, in the case of *l* control channels, can be written as follows:

$$P = \Sigma_{j=1}^{l} S_{yy,jj}^{ref} W_{jj} + 2\Sigma_{j=1}^{l-1}\Sigma_{k=j+1}^{l}\sqrt{\gamma_{jk}^2 S_{yy,jj}^{ref} S_{yy,kk}^{ref}}|W_{jk}|cos\left(\phi_{jk} - \xi_{jk}\right) \quad (1)$$

where

$$S_{yy,jk}^{ref} = |S_{yy,jk}^{ref}|e^{i\phi_{jk}} = \sqrt{\gamma_{jk}^2 S_{yy,jj}^{ref} S_{yy,kk}^{ref}}e^{i\phi_{jk}} \quad (2)$$

and

$$\mathbf{W} = \mathbf{F^H F} \quad (3)$$

in which $\mathbf{S_{yy}^{ref}}$ is the reference SDM, γ_{jk}^2 is the magnitude square coherence between the jth and k^{th} channel, $\mathbf{F}$ is the transfer function matrix between the control and response point, ϕ_{jk} and ξ_{jk} are the phase between the jth and k^{th} terms of the reference SDM and of the $\mathbf{W}$ matrix, respectively. .

For any given set of PSDs, Equation (1) is maximum when the coherences and the cosines are equal to 1. Therefore, to maximize the trace of the response SDM the following condition must be satisfied:

$$P\ is\ maximum\ \Leftrightarrow \begin{cases} \gamma_{jk}^2 = 1 \\ \phi_{jk} = \xi_{jk} \end{cases} \quad \forall j, k = 1 : l, j \neq k \quad (4)$$

Furthermore, the *phase pivoting algorithm* can be used to ensure that the chosen phases determine a physically realizable target [6]. In particular, Equation (4) is applied to the $(l-1)$ terms with the biggest $|W_{jk}|$. The remaining terms are retrieved by respecting the existing phase dependencies $(\phi_{jk} = \phi_{ik} - \phi_{ij})$. Once the EDRM is used to retrieve the phase and coherence of the reference SDM, the *l* reference PSDs required to obtain the *m* responses can be calculated by numerically solving the following set of equations:

$$\begin{cases} S_{xx,11} = \Sigma_{j=1}^{l} S_{yy,jj}^{ref} G_{jj}^{1} + 2\Sigma_{j=1}^{l-1}\Sigma_{k=j+1}^{l}\sqrt{\gamma_{jk}^2 S_{yy,jj}^{ref} S_{yy,kk}^{ref}}G_{jk}^{1} cos\left(\phi_{jk} - \alpha_{jk}\right) \\ \cdots \\ S_{xx,mm} = \Sigma_{j=1}^{l} S_{yy,jj}^{ref} G_{jj}^{m} + 2\Sigma_{j=1}^{l-1}\Sigma_{k=j+1}^{l}\sqrt{\gamma_{jk}^2 S_{yy,jj}^{ref} S_{yy,kk}^{ref}}G_{jk}^{m} cos\left(\phi_{jk} - \alpha_{jk}\right) \end{cases} \quad (5)$$

where $G^m = F(:,m)^H F(m,:))$ and α_{jk} is the phase angle of $G(j,k)$. In Equation (5) matrix G^m can be retrieved during the system identification phase of the MIMO random test, the coherences γ_{jk} and phases ϕ_{jk} are determined using Equation(4), $S_{xx,mm}$ are the target response PSDs on the DUT, and the only unknown terms are the reference PSDs $S_{yy,jj}^{ref}$. The solution of Equation (5) allows to find the minimum reference PSDs that determine the DUT target response. As a consequence, the MPM guarantee the reduction of the drives required to conduct the test while maintaining the response on the DUT constant.

Experimental Verification

The aim of experimental activity is to showcase the applicability and potential of the proposed procedure. The experiments are carried out on a specifically designed specimen, consisting of an cantilever aluminum beam. The main objective of this test case is to demonstrate the ability of the *MPM* to mimic the fatigue damage of a target vibration environment using the least amount of energy possible. The target vibration environment is obtained from a test conducted with flat uncorrelated PSDs as excitation. Furthermore, the responses and the drive PSDs obtained with the target vibration environment and the MPM are compared to those obtained by conducting a test with the same specifications as the target vibration environment but correlated using the *Minimum Drives Method (MDM)*, which is a state of the art drive reduction technique. An additional test is performed by using the *MDM* to directly control the response of specimen to match the target vibration environment. The tests are conducted using the three-axial electrodynamic shaker available at the University of Ferrara. The data acquisition hardware and vibration control software used are the SCADAS Mobile SCM202V (V8 input and DAC4 output modules) and MIMO Random of Simcenter Testlab, respectively.

The specimen chosen for the tests is a cantilever aluminum beam with rectangular section and U-shaped notches near the fixed end. A lumped mass of 0.47 kg is placed on the free-end of the beam, to tune its natural frequencies in a convenient frequency range. The detailed geometry of the beam is depicted in Figure 1.

Figure 2 shows the specimen mounted on the shaker table and the sensors used during the tests. Two accelerometer are used. The *tab* accelerometer is placed near the fixed end of the specimen and the *tip* accelerometer is placed on the lumped mass. The *tip* accelerometer is the location where the vibration environment is measured to obtain the PSDs to be replicated using the *MPM*. Additionally, a strain gauge rosette is bonded near the notches, allowing to study the fatigue damage inflicted to the specimen. The accelerometers are ICP (Integrated Circuit Piezoelectric) triaxial accelerometers by *PCB Piezotronics*, model 356B21. The strain gauge used is a 3/120-RY93 by HBM, which allows to measure a multi-axis strain state. In order to conduct the fatigue analysis, the obtained stress state is converted into an equivalent uni-axial stress state using the Von Mises PSD method.

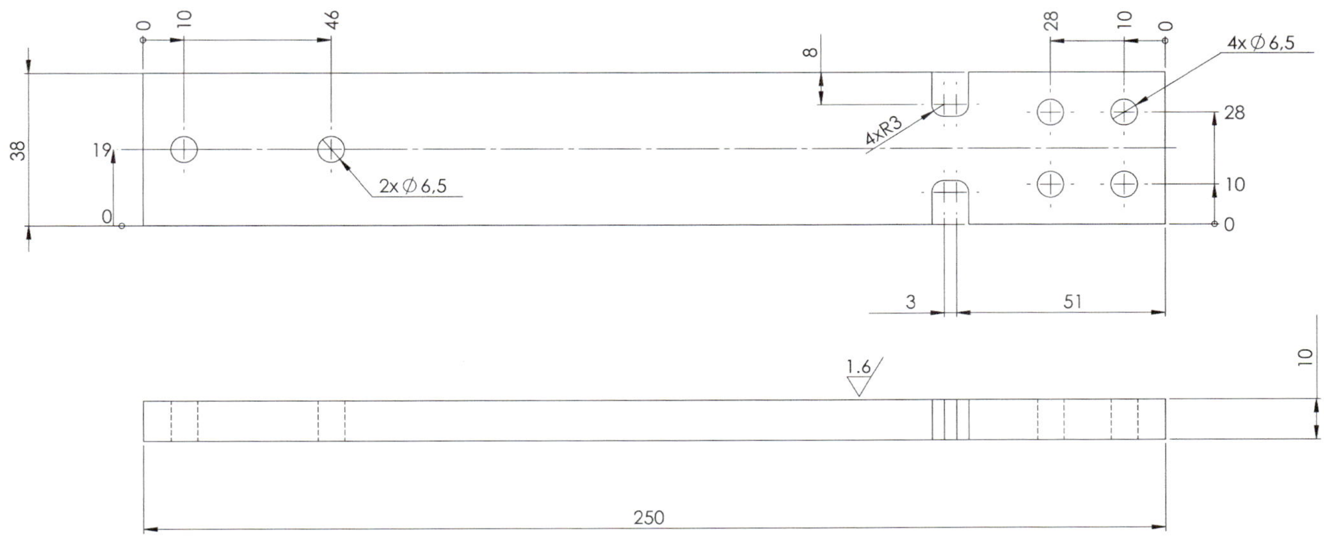

Fig. 1 Detailed geometry of the specimen used.

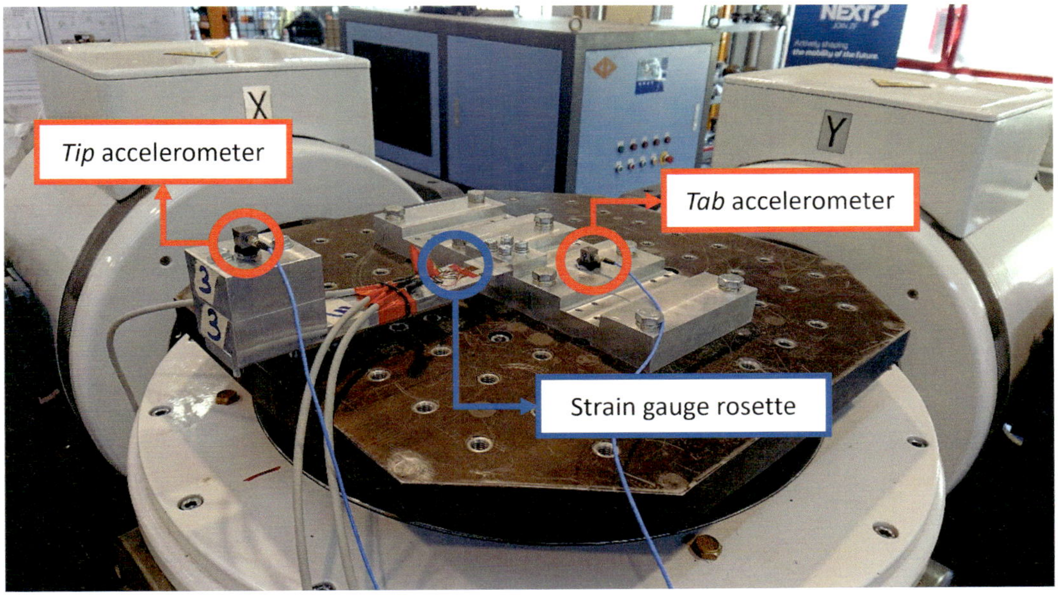

Fig. 2 Specimen mounted on the shaker table and sensors used.

The test specification used to generate the target vibration environment consists of flat uncorrelated profiles of 1 gRMS in the frequency range $[20 - 600]\ Hz$, replicated simultaneously on each direction the *tab* accelerometer.

Figure 3 depicts the comparison of the response PSDs measured the *tab* accelerometer. The three subplots on the left show the acceleration PSDs along the three orthogonal direction measured by the *response* accelerometer. These include the target PSD (blue solid line), the PSD obtained using the *MPM* (orange dashed lines), the PSD obtained in the test performed controlling the *tab* accelerometer and using the *MDM* (yellow dotted-dashed line) and the PSD obtained when controlling the *tip* accelerometer and using the *MDM* (purple dotted line) . On the right side of Figure 3, the bar plot reports the acceleration RMS value measured by the *tip* accelerometer for each direction.

It can be observed that test performed using the *MPM* and the test performed using *MDM* and controlling the *tip* accelerometer (labeled *MDM tip*) both provide a good match with the *target* measurement PSDs. Furthermore, the acceleration RMS obtained by the *MPM* and the *MDM tip* tests is close to the *target* RMS for each direction. In particular, the target acceleration RMS are $8.57g$, $20.01g$ and $5.30g$, the *MPM* test achieved an acceleration RMS of $8.89g$, $20.1g$ and $5.05g$, the *MDM tip* test achieved an acceleration RMS of $8.77g$, $19.71g$ and $5.43g$ along the X, Y and Z axis, respectively. On the other hand, the *MDM tab* test, performed controlling the *tab* accelerometer by using the *MDM* and with the same excitation PSDs of the *target*, determined different results. In particular, the response of the specimen is lower than the response obtained during the *target* measurement. The difference is significant especially along the Y and Z direction where the acceleration RMS determined during the *MDM tab* test is $12.73g$ and $3.77g$, respectively. In fact, changing the correlation of the refer-

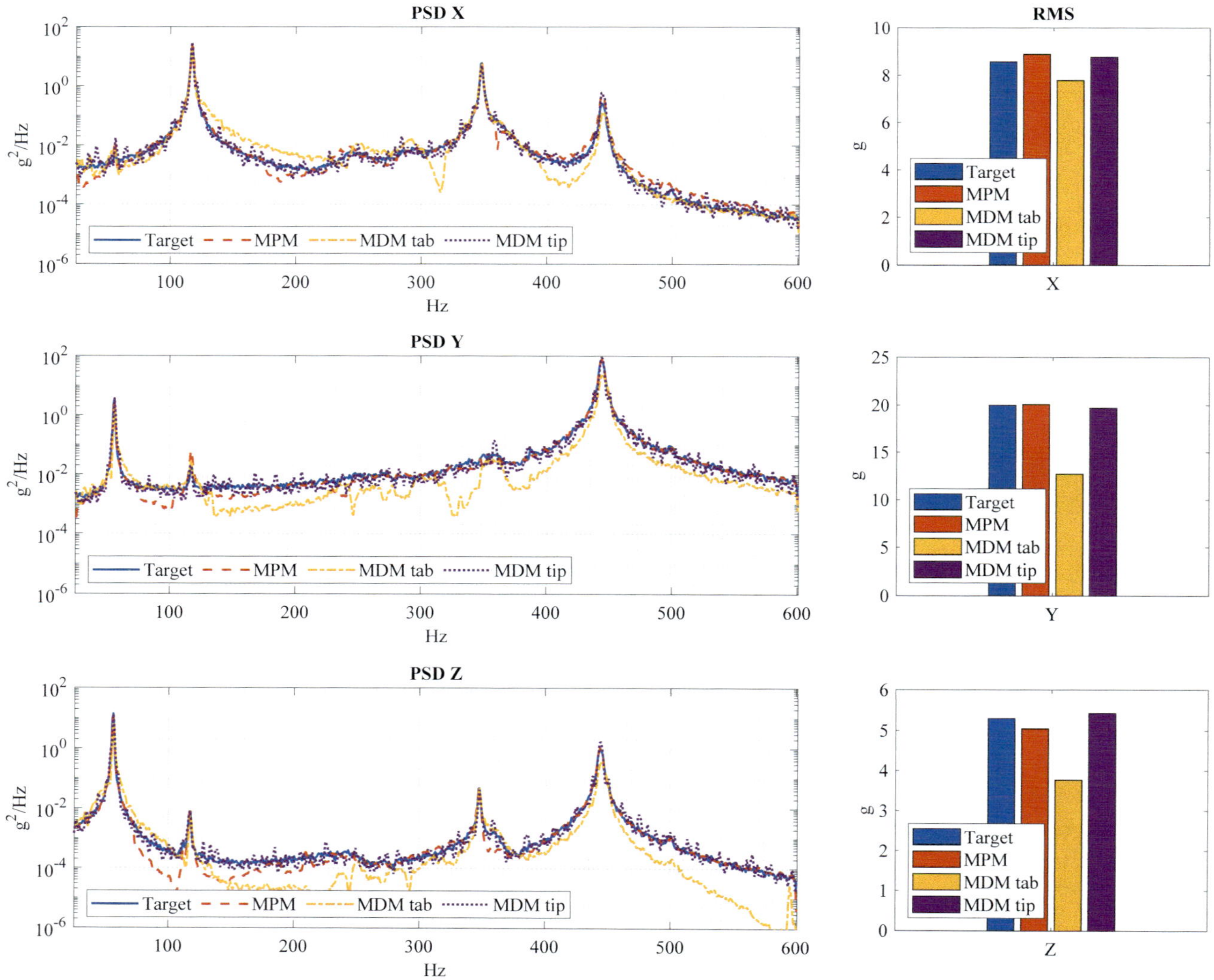

Fig. 3 Acceleration PSDs measured by the *tip* accelerometer during the tests and their respective RMS value

ence SDM can significantly affect the response intensity on the DUT. Therefore, not adjusting the reference PSDs may cause significant deviation from the target vibration environment. These results highlight how the *MPM* is capable of replicating the target vibration environment. Furthermore, Figure 4 presents the comparison between the drives required to conduct each test. The plots on the left show the drive PSDs, the bar plot on the right illustrates the difference in voltage RMS value in dB with respect to the drives required by the *target* measurement. It can be noticed that the *MPM* achieved the greatest drive reduction for each one of the three drives. In particular, the *MPM* achieved a $1.46dB$, $0.89dB$ and $1.02dB$ reduction for drive 1, 2 and 3, respectively. On the other hand, the *MDM tab*, achieved a reduction of $0.23dB$, $0.11dB$ and $0.25dB$ for drive 1, 2 and 3, respectively. It should also be noticed that the *MDM tip* test determined a significant increase in the energy required to conduct the test. This is caused by the physical constraints of the system. In fact, when controlling the response of the specimen, the control target must abide to the dynamic behaviour of the specimen. In particular, at the resonance frequencies the response of the DUT is driven by its natural mode shape and the phase and coherence of the response are independent of the phase and coherence of the multi-axis excitation. Figure 5 shows the comparison between phase of the CSDs obtained during the *MDM tip* test (red crosses) and the phase required to achieve the minimum drives according to the *MDM* when controlling the *tip* accelerometer (green circles) at each natural frequency. The phase obtained during the test is determined by the dynamic behaviour of the specimen and, as a consequence, does not always match the phase required by the *MDM*. This phase mismatch between the control and target phase determines an increase in the required drives. Moreover, at the

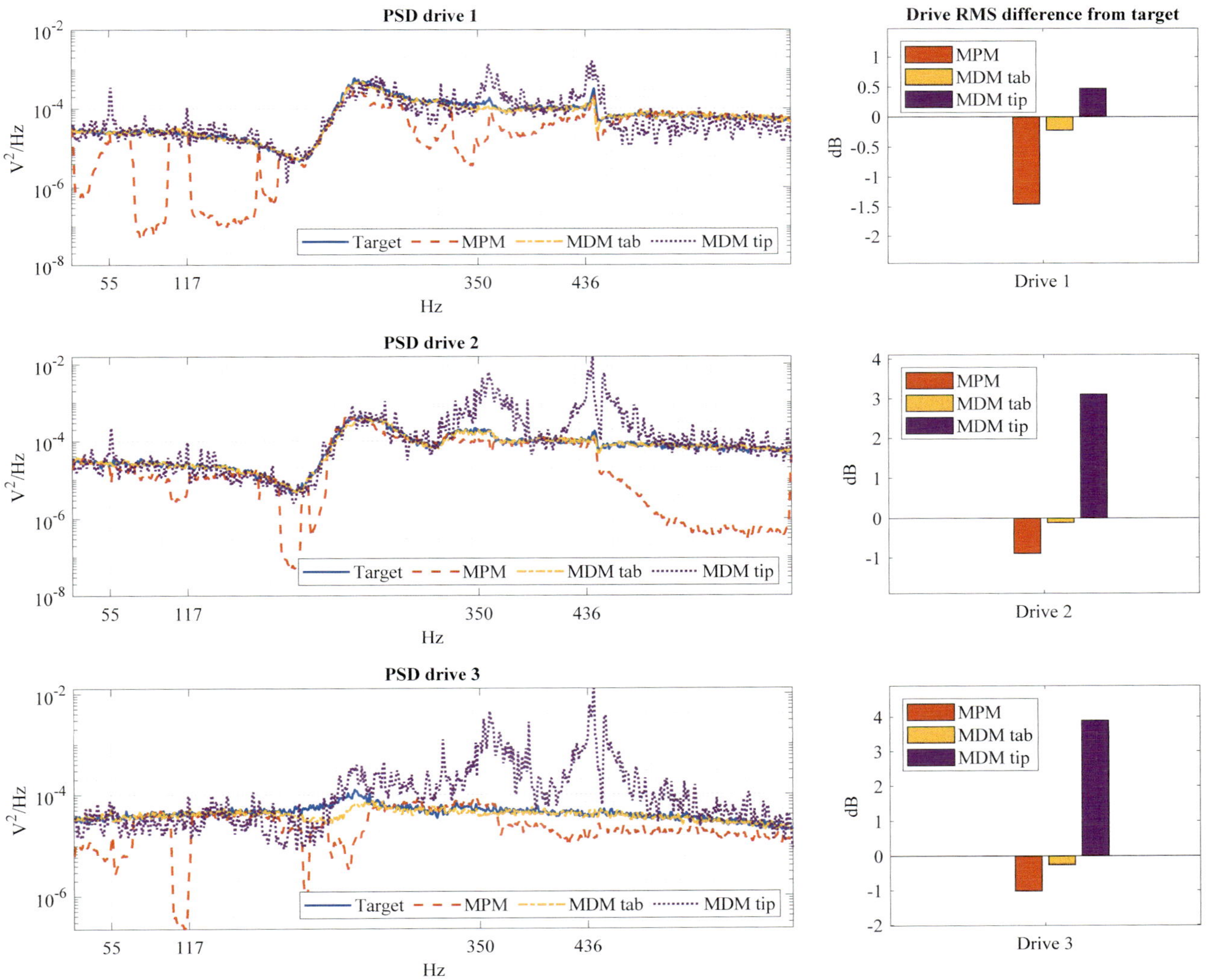

Fig. 4 Comparison of the drives required by the *target measurement*, the *MPM* and the *MDM* when controlling the *tab* and *tab* accelerometer (dB reference: drive 1: 0.23 V, drive 2: 0.22 V, drive 3: 0.16 V).

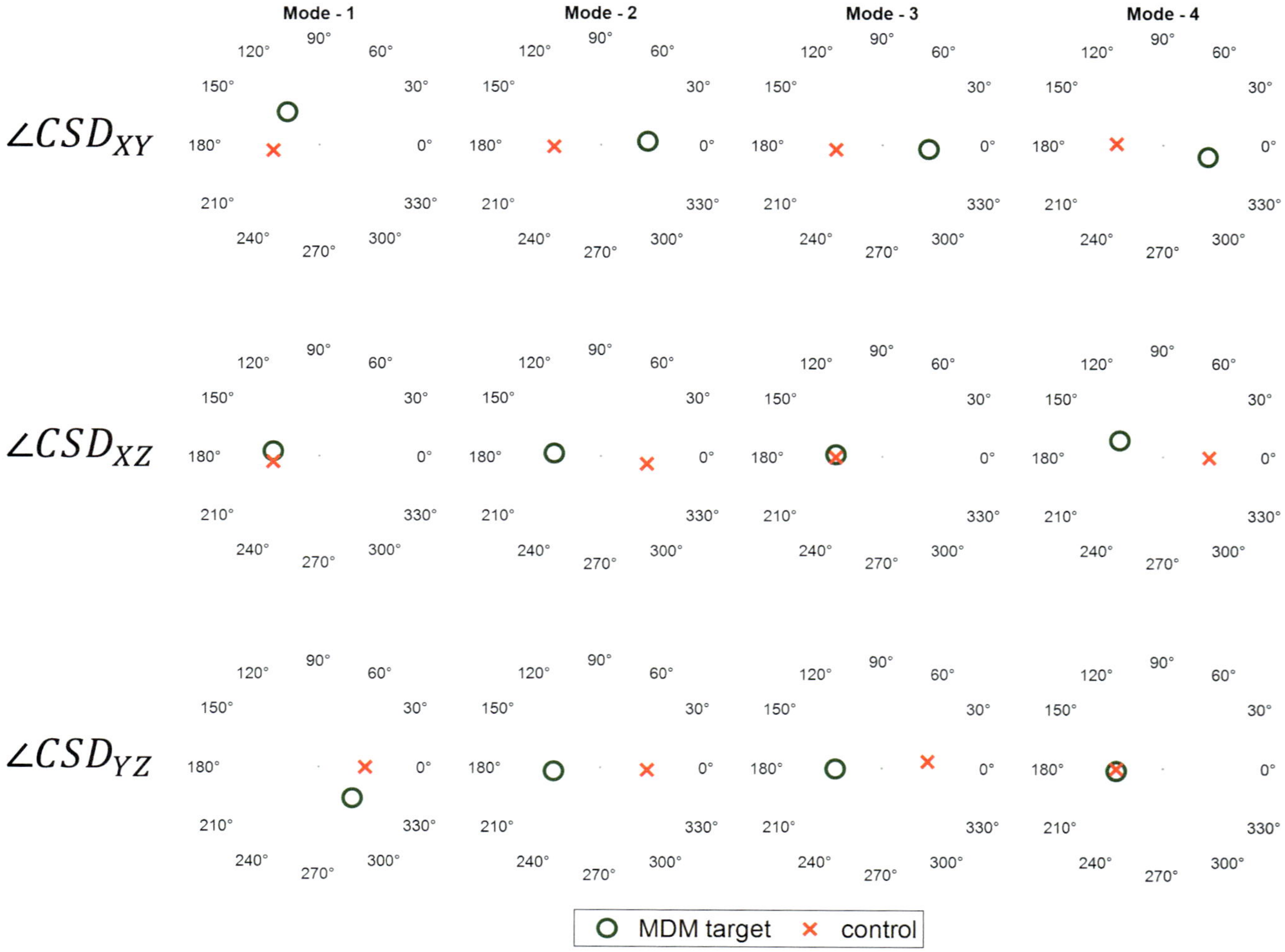

Fig. 5 Phase obtained during the *MDM tip* test on the *tip* accelerometer (red cross) compared to the phase required to achieve the minimum drives (green circles) at each resonant frequency.

second, third and fourth natural frequencies, the phase obtained during the *MDM tip* test are shifted by 180 *deg* with respect to the *MDM* target. As a consequence, the drives are maximized at these frequencies, as displayed in Figure 4.

The last comparison involves the analysis of the fatigue damage inflicted to the specimen by each test. During each test, the PSDs of the strains were measured using the strain gauge rosette placed at the notch of the specimen. The measured strains are then processed to calculate the Von Mises stress PSD experienced by the specimen. Finally, the Dirlik cycle counting method is used to obtained the amplitude probability distribution for each test and compute the fatigue damage [10]. The left plot of Figure 6 shows the amplitude probability distribution calculated from the Von Mises stress PSD obtained during the *target measurement* (solid blue line), the one obtained with the *MPM* (orange dashed line) and the one obtained with the *MDM* when controlling the *tab* accelerometer (yellow dotted-dashed line) and the the *tip* accelerometer (purple dotted line). The bar plot on the right shows the predicted time to failure (TTF) in hours, derived from each probability distribution. It can be noticed that the *MPM* and the *MDM tip* tests determined a probability distribution very similar to the *target measurement*. On the contrary, the *MDM tab* test determined a completely different amplitude probability distribution. As a consequence the predicted TTF of the *MPM* and the *MDM tip* tests are very close to the predicted TTF of the *target*. On the contrary, the *MDM tab* test determined a significant increase in the predicted TTF. This is to be attributed to the reduced response determined by the *MDM*, as depicted in Figure 3. In particular the predicted TTFs are $9.17h$, $10.46h$, $29.89h$ and $9.06h$ for the *target, MPM, MDM tab* and *MDM tip* test, respectively. In conclusion, the results of Figure 3, 4 and 6 highlight the capabilities of the proposed methodology. The *MPM* proved to be able to generate the same responses and inflict the same damage to the specimen as the *target measurement*, while minimizing the drives required to conduct the test. In particular, in comparison to the state of the art solution *MDM*, the *MPM* required lower drives to achieve the same response as the target

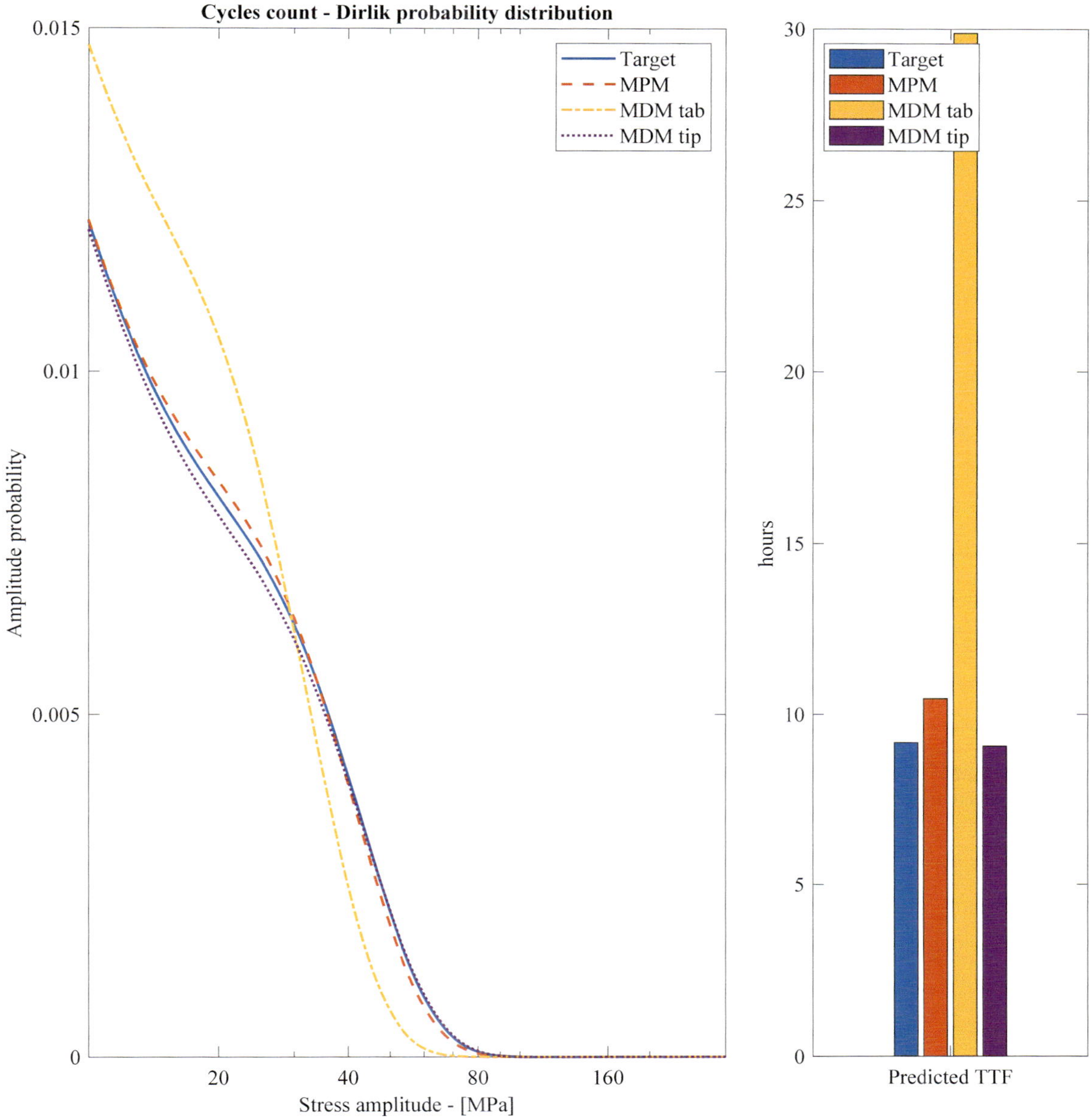

Fig. 6 Amplitude probability distribution (Dirlik) of the Von Mises stress PSDand expected time to failure (TTF) of each test

measurement. Moreover, the *MDM* was not able to simultaneously minimize the drives and maintain the *target* response on the specimen. On the contrary, the *MPM* has completely preserved the damaging effect of the target vibration environment, while demanding the lowest ammount of energy to conduct the test.

Conclusions

This manuscript proposes a novel methodology to define the target for a MIMO random control test, the *Minimum PSDs Method*. The method provides the minimum excitation PSDs to be applied to the DUT that allow to replicate the response of the DUT caused by a different set of excitations while minimizing the voltage required by the shakers to conduct the test. The effectiveness of the *MPM* has been verified through a dedicated experiment. Moreover, the *MPM* has been compared to the *MDM*, a state of the art solution for drive power reduction in MIMO random tests.

The *MPM* determined the highest reduction of the drives, while preserving the response and the fatigue damage induced in the specimen by the *target*. On the contrary, when controlling the *tab* accelerometer, the *MDM* required higher drives the *MPM* and determined a different response of the specimen which lead to a significant under-testing of the DUT. Moreover, when controlling the response of the DUT to match the *target* responses, the *MDM* was able to replicate the *target* responses. However, the drive minimization was not successful because of the physical constrains of the DUT, which lead to an increase in the required voltage. It can be concluded that the *Minimum PSDs Method* allows to replicate a set of responses on the DUT while reducing the total power required to conduct the test. The methodology may allow to re-design multi-axis random control tests, reducing the drive requirements while preserving the damaging effects of a targeted vibration environment.

References

1. Underwood, M.A. “Multi-exciter testing applications: theory and practice”. *Proceedings-institute of environmental sciences and technology*, pages 1–10 (2002)
2. Peeters, B. and Debille, J. “Multiple-input-multiple-output random vibration control: Theory and practice”. *Proceedings of the 2002 International Conference on Noise and Vibration Engineering, ISMA*, page 507 – 516 (2002)
3. Smallwood, D.O. “Multiple-input multiple-output (mimo) linear systems extreme inputs/outputs”. 14(2):107 – 131 (2007)
4. Smallwood, D. “A proposed method to generate a spectral density matrix for a multiple input, multiple output (mimo) vibration test”. *Proc. of 80th Shock and Vibration Symposium.*
5. D’Elia, G., Musella, U., Mucchi, E., Guillaume, P., and Peeters, B. “Analyses of drives power reduction techniques for multi-axis random vibration control tests”. *Mechanical Systems and Signal Processing*, 135:106395 (2020)
6. Musella, U., D’Elia, G., Carrella, A., Peeters, B., Mucchi, E., Marulo, F., and Guillaume, P. “A minimum drives automatic target definition procedure for multi-axis random control testing”. *Mechanical Systems and Signal Processing*, 107:452–468 (2018)
7. D’Elia, G., Mucchi, E., and Dalpiaz, G. “A novel methodology for dynamic response maximisation in multi-axis accelerated random fatigue testing”. *Mechanical Systems and Signal Processing*, 181 (2022)
8. D’Elia, G., Proner, E., and Mucchi, E. “Accelerated random fatigue testing by means of a triaxial electrodynamic shaker: solutions for combining the multiple test specifications”. page 1224 – 1232 (2022)
9. Zheng, R., Xie, W., Wei, X., and Chen, H. “Limit strategy for multi-input multi-output random vibration control”. *Mechanical Systems and Signal Processing*, 195 (2023)
10. Dirlik, T. “Application of computers in fatigue analysis.”. *Ph.D. Thesis, University of Warwick* (1985)

Chapter 6
Making Modal Analysis Easy and More Reliable – Challenging Ai-Based Algorithms with the Barc Example

Tim Kamper, Denis Beljan, Patrick Hüskens, and Haiko Brücher

Abstract Though modal analysis is a common tool to evaluate the dynamic properties of a structure, there are still many individual decisions to be made during the process which are often based on experience and make it difficult for occasional users to gain reliable and correct results.

The paper presents how on different steps of the process experience-based decisions can be supported and replaced by automated evaluations to make modal analysis accessible to less experienced users. In addition to traditional methods such as the Least-Squares Complex Frequency-domain (LSCF) estimator the presented approach takes advantage of more innovative methods such as a neural network to reduce experience-based decisions and increase reliability.

The advantage of the presented approach is shown based on the example of a Box Assembly with Removable Component (BARC). BARC is known as a challenge structure for approaches for the evaluation of dynamic properties. In the study two groups of users are exposed to perform a modal analysis of the BARC. The groups are selected as heterogeneous groups with different levels of experience in modal analysis. The first group performs the modal analysis based on classic algorithms and methods whereas the second group is supported by AI based technology.

Keywords Modal Analysis · Optimal Driving Point · AI

Introduction

Modal analysis is a widely used method to analyze the dynamic properties of structures. It is used for a variety of tasks such as troubleshooting, validation of numerical models and the parametrization of numerical models. Due to its various use cases the user group of modal analysis is very heterogeneous with different background-knowledge and experience. Unfortunately, the process of modal analysis, that starts the planning of the measurements and ends with the postprocessing of the results requires a lot of decisions based on the experience of the user. Thus, it is difficult to judge on the reliability of the results of modal analysis and the whole process is hardly reproducible. Furthermore, wrong decisions at an early step of the process can often only be discovered in the very end of the process, if at all. This is particularly painful, as the process of modal analysis can be a quite time-consuming task.

This paper focuses on the evaluation of different methods that are designed to contribute to an easier process of modal analysis that delivers more reliable results.

Figure 1 gives an overview of the process steps of modal analysis with the corresponding decisions that are necessary to be made by the individual user. All these decisions have an impact on the completeness and the quality of the results, such as number of extracted mode shapes, accuracy of the fitted frequency response functions (FRF) and of the detected mode shapes. Before the user can start the experimental data acquisition, the test must be designed properly. During the test design phase, the used data acquisition method and hardware must be chosen (shaker excitation, impact testing with roving hammer/accelerometer, laser vibrometer, etc.). Choosing the right geometric resolution of measurement points enables capturing local shape characteristics. The selection of correct number and positions of good reference degrees of freedom (DOFs) is crucial, as it results in satisfactory signal-to-noise-ratio (SNR) and deflection magnitudes for all modes within the desired frequency range, enabling the detection of all corresponding eigenfrequencies during postprocessing. Furthermore, the number of reference DOFs may be kept down by a good selection. This leads to less equipment needed and less data to

Tim Kamper · Denis Beljan · Patrick Hüskens · Haiko Brücher
HEAD acoustics GmbH, Ebertstr. 30 a, D-52134 Herzogenrath, Germany
e-mail: Tim.Kamper@head-acoustics.com; Denis.Beljan@head-acoustics.com; patrick.hueskens@head-acoustics.com; haiko.bruecher@head-acoustics.com

Alexandra Karlicek et al. (eds.), *Dynamic Environments Testing, Vol. 7*, Conference Proceedings of the Society for Experimental Mechanics Series, https://doi.org/10.13052/9788743801528-6

be acquired and processed. During the step of data acquisition, the user must configure settings for the frequency response functions (FRFs) like the time to be recorded, block size, windowing of excitation and/or response. Sensor ranges must be adapted to induce a good SNR without overloading. Depending on the structure, there is a bad reproducibility at some measurement points (i.e. non-linear behavior) and in the case of impact testing, the user might be confronted with double hits. Once the data acquisition is complete, the modal parameters can be extracted in the next step wherein the curve fitting method must be selected. The next experience driven decision is to further parameterize the curve fitting method itself. The method outputs pole candidates which then must be grouped, selected and extracted by the user. Finally, the modal parameters consisting of mode shapes with corresponding eigenfrequencies, and damping values form the modal model which can be evaluated in the postprocessing step.

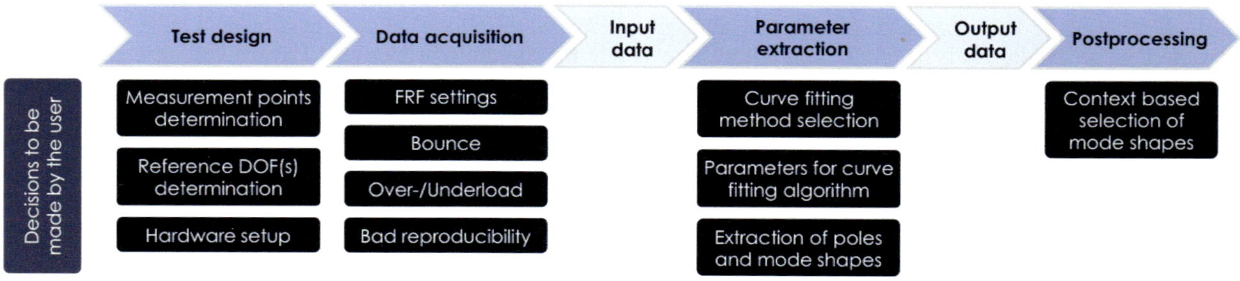

Fig. 1 Process of experimental modal analysis and steps that require user interaction, affecting result completeness and quality.

This paper focuses on the validation of solutions for two different steps in the process of modal analysis: The initial step of test design for the modal analysis and the step of parameter extraction where the user must provide input parameters for the curve fitting algorithm. Both steps and the corresponding decisions to be made by the user will be described in the following sections. Moreover, the suggested solution approaches based on AI will be discussed.

To validate the suggested solution approaches a study with participants that have different experience levels regarding modal analysis was performed. The study demonstrated the advantages of the suggested approaches and will be discussed in the last section of this paper.

Identified Problems and Solution Approaches

Test design: Reference DOFs identification

In the step of test design, the user must decide among other on the correct number and positions of reference DOFs. The correct number and positions of reference DOFs is important to make sure that every eigenfrequency present in the frequency range of interest can be detected. This is particularly critical because a missing eigenfrequency, and thus an incomplete modal model, can hardly be detected by the user.

The importance of the correct choice of reference DOFs can be illustrated by the simplified examples given in Figure 2, where the first modes of two simple oscillating structures are depicted. If the reference DOF is placed at the position marked in red, the respective mode cannot be detected because the reference DOF is placed in a node of the mode shape.

Placing the reference DOF at or near a node of the mode shape is not suitable for introducing energy into the structure to excite the specific frequency if used as an excitation DOF, nor for detecting a vibrational amplitude if used as a measurement DOF.

Using simple examples like they are given in Figure 2 the decision on the number and positions of reference DOF is simple and straight forward. But for real, complex structures with a lot of modes this decision is not easy and cannot be made without any previous thoughts and analyses.

For the identification of the number and positions of reference DOFs different methods exist. They can be divided into two groups by the fact whether a numerical model is needed to get insights into the dynamic properties of the structure or not.

- Methods based on numerical simulations: Based on the modal deformation derived from numerical simulations it is possible to decide on the number and positions of reference DOFs [1, 2]. In [3] a method was introduced which works based on a non-validated numerical model in a fully automated manner.

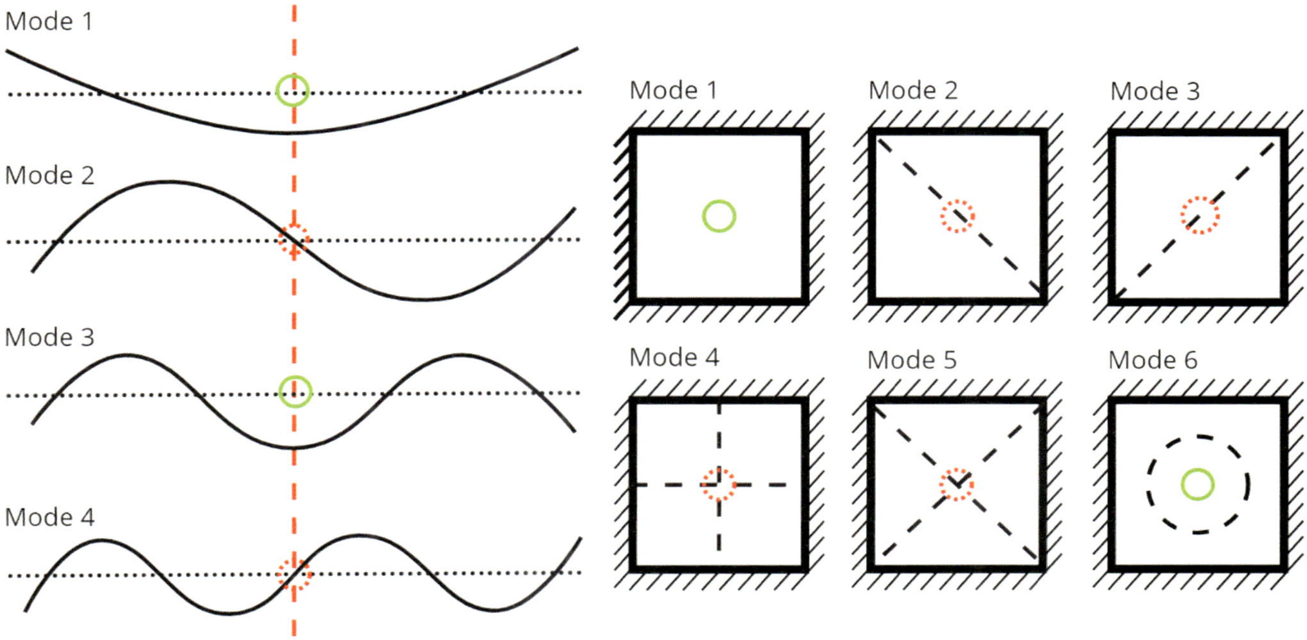

Fig. 2 Examples for the importance of a good choice of the reference DOF (red circle: mode cannot be detected, green circle: mode can be detected) [4].

- Methods based on tests: When there is no numerical model available the decision on the optimal number and positions of reference DOFs is much more difficult. Since there is no prior knowledge of the structure available the decision must be made based on measurements. The simplest and most common approach is the visual inspection of a few driving point measurements (DPM) at different potential reference DOFs. To perform a driving point measurement, an accelerometer is placed at one potential reference DOF and the structure is excited as close to this point as possible using an impact hammer. By visual inspection of the measured DPM and based on his or her individual experience the user must decide what are the modal properties of the structure under test and which of the potential reference DOFs is suited best to detect all the modes of interest. For simple structures with a limited number of frequencies in the frequency band of interest, this method may work, but it is hardly reproducible, and the effort raises with the complexity of the structure.

In this study we focus on the validation of a method which was introduced in [4] and allows to identify the optimal number and positions of reference DOFs without prior knowledge or experience based on measurements in an automated manner. In the following a brief description of the method will be given. For more details, please refer to the corresponding paper [4].

Figure 3 gives an overview of the steps in the process of the test based prestudy.

Before the start of the prestudy the frequency range of interest must be defined. The study then aims to find the optimal number and positions of reference DOFs to achieve a complete modal model containing all the modes in the respective range of interest. The prestudy relies on driving point measurements at potential reference DOFs. Since no prior knowledge is available, with every DPM added to the prestudy the knowledge of the dynamic properties and hence the eigenfrequencies to be detected increases. One core element of the prestudy is the automated detection of poles for every driving point measurement which is done with the help of a neural network. The following step is the processing of the detected poles for the whole dataset. This step can be separated into three subtasks:

- First all the poles detected in different DPMs are captured in one group.
- In a second subtask the dynamic properties of the structure (eigenfrequencies, damping) known so far are derived from the group of poles.
- In the third subtask it is analyzed which pole is suitable to detect which frequency.

The outcome of the step of processing the detected poles for the whole dataset is a list of the DOFs which were tested until the current state including the information which DOF is suitable to detect which eigenfrequency. This list is then used

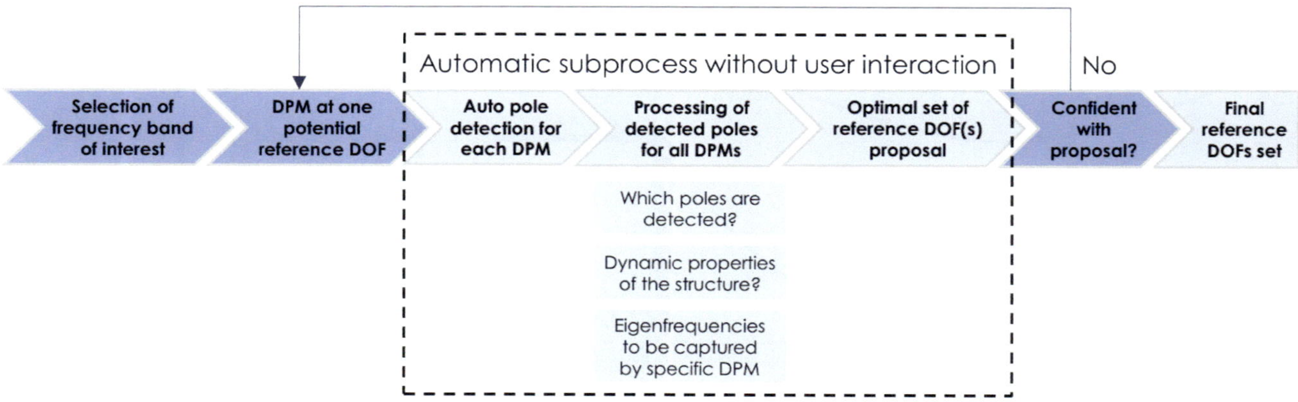

Fig. 3 Process of test based prestudy.

in a final step to generate a suggestion of the minimal number of reference DOFs which are needed to make sure that every eigenfrequency can be detected in the consecutive modal analysis. The user than can decide whether he/she is confident with the suggestion. If not an additional DPM must be measured and be fed to the algorithm and the process starts all over again. It is important to notice that besides the conduction of the driving point measurement the whole process works in an automated manner to relieve the user. Since no prior knowledge of the structure is available, the algorithm learns about the structure during the conduction of the prestudy and with every measurement taken, the knowledge about the dynamic properties of the structure is more and more complete. Since there is no reference against which to compare the identified eigenfrequencies, there is no hard criteria for the end of the study. Instead, the study can be regarded as complete, when the number of identified modes and the number and positions of suggested reference DOFs does not change for a few iterations.

Depending on the chosen method for modal analysis (roving excitation or roving accelerometer) the identified reference DOFs are then declared as excitation DOFs or measurement DOFs and the next steps according to Figure 1 can be tackled.

Input data: Parameterization of parameter extraction

For the step of parameter extraction displayed in Figure 1 different curve fitting algorithms have been developed in the past. The curve fitting algorithms aim to extract the modal properties of the structure from the measured data. The Least Squares Complex Frequency Method (LSCF-Method) [5] is currently one of the most common methods and is widely used throughout the modal analysis community. The LSCF-Method needs to be parameterized by the user to identify all poles existent in the dataset. A wrong choice of parameters leads to incomplete results or results of poor quality. Since the LSCF-Method iteratively attempts to fit a mathematical model of increasing order to the measured data, the highest requested order must be defined by the user. If the highest order is too low, then the LSCF-Method tends to miss poles and consequently the derived modal model is incomplete, leading to the disadvantages discussed above. If the highest order is too high, then the LSCF-Method tends to identify spurious poles which confuse the user and mess up the further analysis of the problem. Furthermore, a too high model order leads to an unnecessarily increased computational effort. The recommended modal order is linked to the number of expected poles in the dataset and the user decides based on his or her experience often in an iterative manner what model order to set as an input parameter for the LSCF. To make this step more reproducible, more reliable and faster in [6] a neural network was introduced which is capable to estimate the number of poles based on measured transfer functions which serve as the input data for the extraction of modal parameters.

Since this paper focuses on the validation of the benefit of the approach for users of modal analysis, at this point only a short description will be given. For more details, please refer to the corresponding paper [6].

The network architecture chosen to perform the task of pole estimation is a convolutional neural network based on convolutional layers. Figure 4 shows the architecture which is, based on its shape also called a U-Net. The left part is known as the encoder, where the neural network reduces the spatial information of the input while extracting the features. The right part is called the decoder, where the feature information is mapped back onto the original spatial representation. In this case the input is a transfer function (FRF) with a fixed number of frequency steps in the format of real part and imaginary part. The output is a vector of the same length as the input and for every frequency step the probability is given that the FRF has a pole at this specific frequency step. From the output of the neural network for all the FRF measured for the respective modal analysis the requested maximal model order is derived and given to the LSCF as an input parameter.

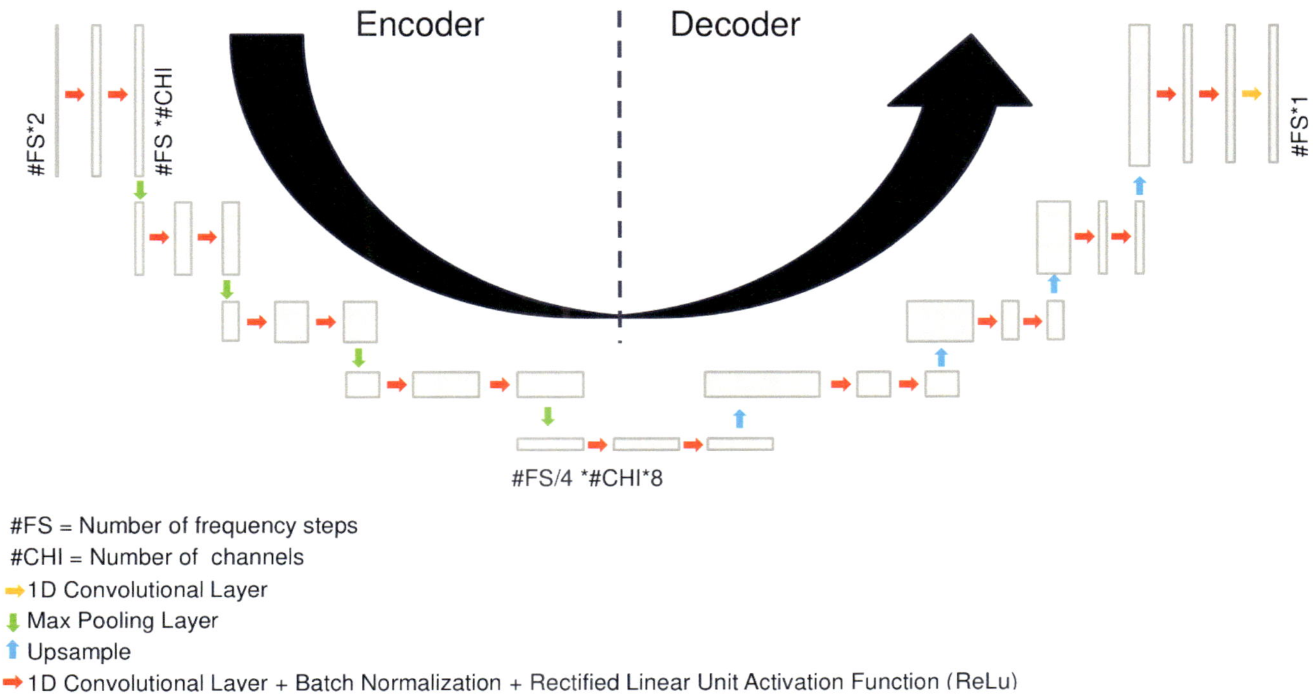

Fig. 4 Network architecture [6].

For the training of the neural network aprox. 700000 datasets with known number and positions of poles where necessary. This is almost impossible to achieve with data from real measurements. This is why the training of the neural network is based on synthesized FRF, which can be produced in an unlimited number with known modal content.

For the synthesis of the FRF an equation known from the curve fitting was used [7]:

$$H_{pq}(\omega) = \sum_{r=1}^{N_m} \left(\frac{A_{pqr}}{j\omega - \lambda_r} + \frac{A^*_{pqr}}{j\omega - \lambda^*_r} \right) \tag{1}$$

In equation 1 the FRF from DOF p to DOF q in dependency of the angular frequency ω is represented by H_{pq}. λ_r and $\lambda_r{}^*$ represent a pole and its conjugate complex value while N_m denotes the number of poles. A_{pqr} and $A_{pqr}{}^*$ denote the residue and its complex conjugate and can be calculated according to equation 2:

$$A_{pqr} = \frac{1}{j2\omega_{dr} m_r} \psi_{pr} \psi_{qr} \tag{2}$$

In equation 2 ψ_{qr} and ψ_{pr} denote the shape coefficients, ω_{dr} is the damped angular eigenfrequency, and m_r represents the vibrational mass of the specific mode.

With the help of the neural network, the step of parameter extraction can be performed significantly faster and is much less error prone.

Validation of Added Value

To validate the added value of the two solution approaches discussed above a study with 19 participants with different background knowledge regarding modal analysis was conducted. The participants were divided into two groups, one had the task to perform a modal analysis without these solution approaches and the other group had the task to perform a modal analysis with the help of these solution approaches. Figure 5 gives an overview of the experience level of the participants in the different groups regarding experimental modal analysis (EMA). It demonstrates that both groups include participants that have never conducted an EMA on their own and have only theoretical knowledge of the process as well as participants who classify themselves as very experienced users.

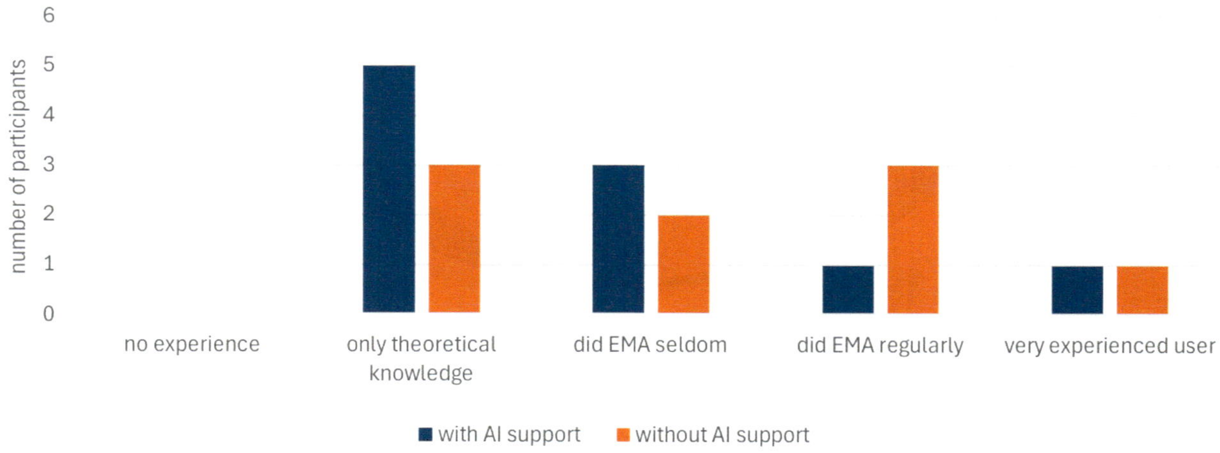

Fig. 5 Experience level of the participants of the study regarding experimental modal analysis (EMA).

From the 19 participants 17 where students, research assistants and professors from the RWTH Aachen University, the University of Applied Sciences Aachen, and the University of Applied Sciences Dortmund. Two participants where employes of HEAD acoustics GmbH.

The object under investigation is the Box Assembly with Removable Component (BARC) structure (see Figure 6). It was designed in 2018 facing the "Boundary Condition for Component Qualification" round robin test setup challenge [8]. Since then, the BARC structure is well known by structural dynamics experts in the American region as it was subject of several investigations and studies, i.e. [9, 10]. It has two main parts, the box and the removable component, which may be coupled either by bolted joints or glue. The box is a symmetric square profile which is cut off in the upper center. The component consists of two pieces of a C-channel, which are connected to a beam in an asymmetric manner. Both main parts are made of aluminum. The simplicity of the parts allows for easy and cost-effective reconstruction. Furthermore, it allows for easy numerical model setup. Due to the asymmetry of the assembled BARC structure, the mode shapes are not easy to predict intuitively even by experts, what makes the selection of appropriate reference points a complicated and potentially time-consuming task.

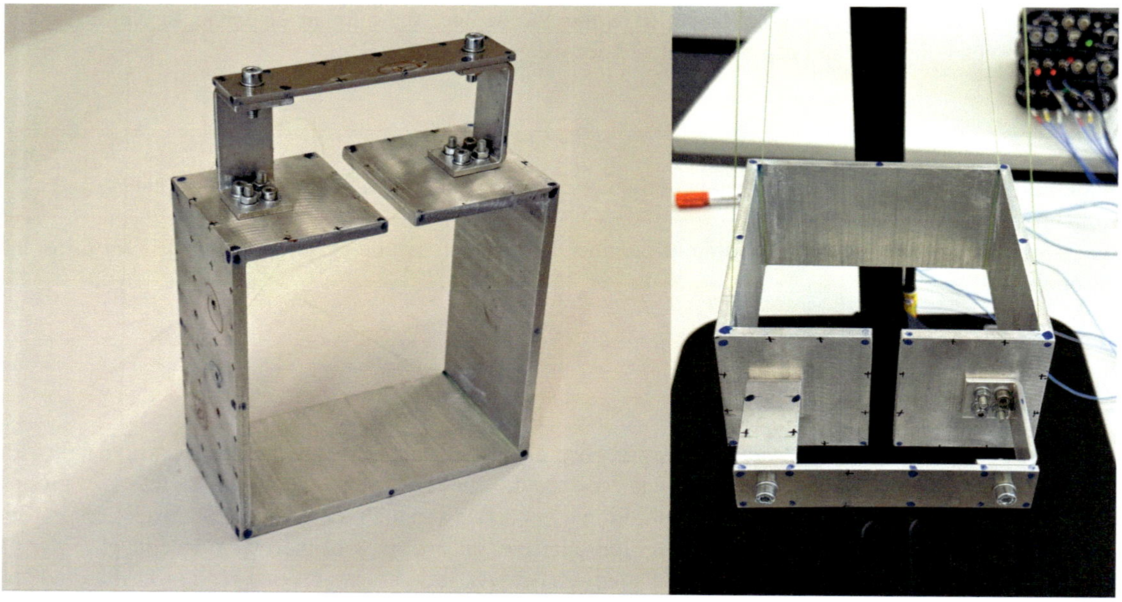

Fig. 6 DUT (left) and suspension of the DUT (right).

To make the study fair and comparable, different measures were taken in advance:

- The DUT was suspended by four threads attached to the corners of the box (see Figure 6). This suspension was kept equal for all participants.
- The method of roving excitation was used throughout the experiments to minimize the effect of mass loading
- The experiments were conducted under the guidance of a supervisor who has experience in both: experimental modal analysis as well as the software in use.
- Same geometrical resolution of excitation points.
  - The positions of the excitation points were predefined by the supervisor.
- All participants used the same type of hardware (e.g. impact hammer, hammer tip, acceleration sensors, frontend).
- The boundary conditions (e.g. frequency range of interest, sampling rate, windowing) were predefined by the supervisor and kept equal throughout the experiments.
- The supervisor gave a short introduction to every participant on the importance of reference point identification and the relevance of the model size for the step of parameter extraction.
- Every participant conducted the EMA separately at a different time. The results of the EMA where not shared with the participants during the ongoing study. Neither a statement regarding the correctness of the results was given.
- The supervisor kept track of time and allowed a maximum of 30 minutes for the choice of reference points as well as a maximum of 30 minutes for the curve fitting.
- The impact measurements itself were conducted by the supervisor to ensure only the approaches discussed above contribute to the results of the participants.

To establish a reference against which to compare the results of the respective participant's modal analysis, the BARC structure was analyzed thoroughly with the help of experimental and numerical modal analysis.

To validate the completeness of the received modal models their similarity to the reference model was analyzed using the Modal Assurance Criterion (MAC)-Matrix. The MAC value is a measure to compare the eigenvectors and thus the mode shapes of different modal models [11]. It takes values between 1 and 0. While a MAC value of 1 means the compared shapes are identical, a value close to 0 means the shapes have no similarity. Thus, the MAC value can be used to verify that the modal model is comparable to the reference model and contains all the desired modes.

Figure 7 displays the results of the conducted study: while all participants in the group which was allowed to use AI support where able to achieve a complete modal model in the frequency range of interest regardless of their individual experience in the field of modal analysis, the majority of participants was not able to achieve a complete modal model without the support of the AI driven tools introduced in the previous chapters. In this study a complete modal model is defined as a modal model which contains all the modes in the respective frequency range but no additional modes which have no physical meaning.

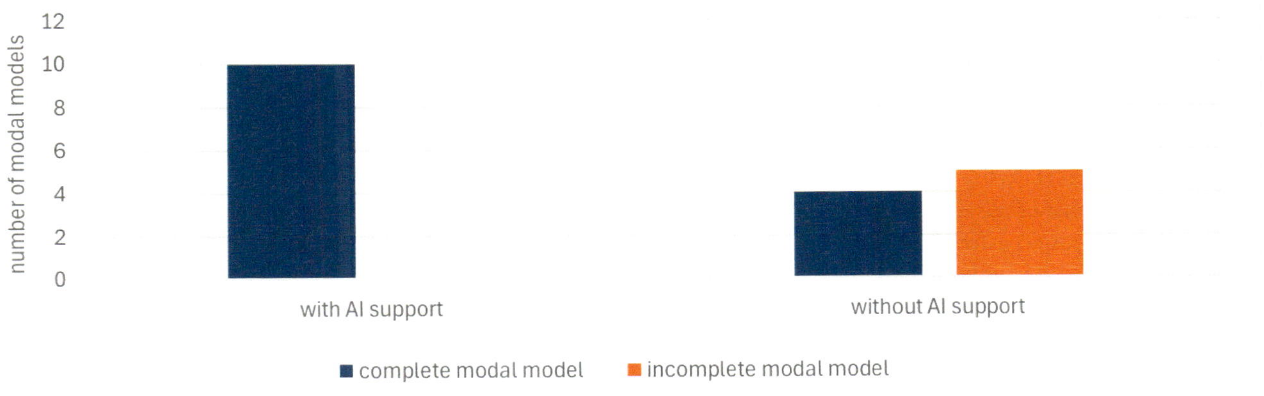

Fig. 7 Number of complete modal models for the groups with and without AI support.

Figure 8 focuses on the group which was not supported by AI approaches and displays the number of complete and incomplete modal models versus the experience level of the participants. Like one would assume, the more experience the participant has the bigger is the probability that he/she can achieve a complete modal model. On the other hand, users with

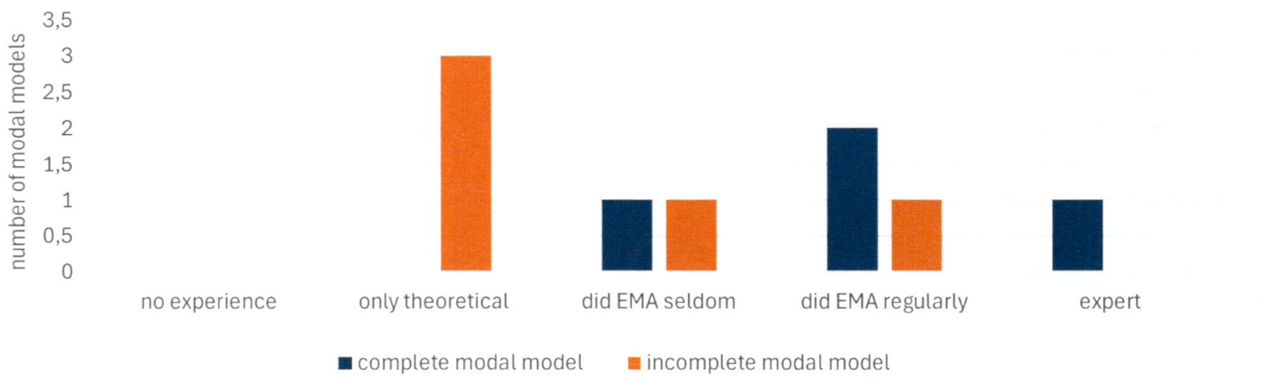

Fig. 8 Results of the study with focus on the group of participants without AI-support.

only theoretical knowledge of the modal analysis had no chance to achieve a complete modal model without the solution approaches presented in this paper. This evaluation is plausible and supports the general result of the study.

Two more remarkable outcomes of the study:

- Two participants using the AI support managed to finalize the step of parameter extraction and curve fitting in the timespan of just one minute. This shows that with the parametrization of the LSCF via the neural network parameter extraction is possible very fast and almost automated.
- The average confidence with the resulting modal model was similar for both, the supported as well es the unsupported participants. The participants were asked to rate their confidence with the results they gained from 1 to 10 with 10 marking the highest confidence and 1 the lowest. The fact, that both groups have a similar average confidence of approx. 7, strengthens the point, which was made in the beginning, that possible mistakes in the process of modal analysis which affect the completeness of the modal model are hard to detect. Without a reference against which to compare these mistakes even might be unrevealed, except that the modal model derived is not suitable to solve the problem it was intended to solve.

Conclusion

In this paper two of many difficulties in the process of modal analysis are addressed: The test design step, where number and positions of reference DOFs must be defined and the step of parameter extraction where the extraction algorithm needs to be parametrized. Both steps are highly dependent on the user's knowledge and experience making it hard for users with less expertise or occasional users to gain reliable end reproducible results of high quality. But even users with more experience in modal analysis tend to spend a lot of thoughts and time on these steps. To solve this issue two approaches are discussed which were introduced in [4, 6] and assist the user among other with the help of AI to gain reliable results independently from their individual experience.

The added value of the two approaches was validated by a study on the example of the BARC structure. For the study two groups of participants performed a modal analysis of the BARC structure one without any support and the other was allowed to benefit from the AI supported approaches introduced in this paper. The study proves that these approaches help users of modal analysis to gain reliable results fast and independent from their individual experience even though no prior knowledge of the structure (e.g. numerical model) is available. This marks another important step towards an easy and reliable process of modal analysis.

References

1. Ewins, D.J., "Modal testing. Theory and practice. 2nd ed. Research Studies Press, Baldock", UK, 2000
2. Wegerhoff, M., "Methodik zur numerischen NVH Analyse eines elektrifizierten PKW Antriebsstrangs", Dissertation, Mainz Verlag, Aachen, 2017

3. Brücher, H., Wegerhoff, M., Sottek, R., "Numerisches Vorstudientool zur Planung der experimentellen Modalanalyse", DAGA 2021
4. Kamper, T., Beljan, D., Brücher, H., and Wegerhoff, M., "Making Modal Analysis Easy and More Reliable – Reference Points Identification by Experimental Prestudy", SAE Technical Paper 2024-01-2931, 2024, doi:10.4271/2024-01-2931
5. Guillaume, P., Verboven, P., Vanlanduit, S., Van der Auweraer, H., Peeters, B., "A poly-reference implementation of the least-squares-complex-frequency-domain estimator", Proceedings of IMAC 2003
6. Kamper, T., Wegerhoff, M., Lobato, T., Sottek, R., "Simplifying the Extraction of Modal Parameters from Test Data using a Neural Network", DAGA 2022
7. Brandt, A., "Noise and Vibration Analysis. Signal Analysis and Experimental Procedures", John Wiley Sons, Chichester, 2011
8. Soine, D. E., Jones, R. J., Harvie, J. M., Skousen, T. J., Schoenherr, T. F., "Designing hardware for the boundary condition round robin challenge", in Topics in Modal Analysis & Testing (M. Mains and B. Dilworth, eds.), vol. 9 of Conference Proceedings of the Society for Experimental Mechanics Series, (Cham), pp. 119-126, Springer, 2019.
9. Rohe, D. P., Schultz, R. A., Schoenherr, T. F., Skousen, T. J., "Comparison of multi-axis testing of the barc structure with varying boundary conditions", in Proceedings of the 37th International Modal Analysis Conference, (Orlando, FL), Jan. 2019.
10. DeClerck, J., Blough, J., Larsen, W., VanKarsen, C., Soine, D., Jones, R., "Sensitivity study of BARC assembly", in Proceedings of the 37th International Modal Analysis Conference, (Orlando, FL), Jan. 2019
11. Allemang, R., The Modal Assurance Criterion (MAC): "Twenty Years of Use and Abuse", In: Sound and Vibration. 37 (1), 2003

Chapter 7
An Undamped Dynamic Vibration Absorber on a Resonant Plate Shock Test

David E. Soine, Adam J. Bouma, and Forrest J. Arnold

Abstract Dynamic vibration absorbers are widely used for vibration reduction on vehicles, buildings, industrial equipment, and many other applications. If an undamped or very lightly damped absorber is tuned to suppress the resonant response of a structure, two new resonances are created at frequencies above and below the suppressed resonance peak. An experimental investigation was conducted to explore the effect of such a vibration absorber on shock response spectra obtained during resonant plate shock testing. The shock response spectrum at the original plate resonant frequency was reduced. The response of the two new resonant frequencies appeared equal in a Fourier spectrum, but only the response of the higher frequency resonance had a strong effect on the shock response spectrum, creating a new "knee frequency" at a higher frequency than that of the original unmodified plate.

Keywords Resonant · Fixture · Plate · Pyroshock · Absorber

Introduction

Resonant fixture shock tests are a means to perform mid-field and far-field pyroshock simulation in the test laboratory. The method can be adapted to simulate aspects of stage separation, ejection shock, blast events, flight shock, and margin tests for other shock environments ordinarily performed on a vibration shaker but outside the shaker test capability. Resonant fixtures in use include square, rectangular, or round plates, beams, and bars of various desirable responding frequencies. Shock practitioners have developed means to adjust the natural frequency and damping of these resonant shock test systems and are on the lookout for new tools to adjust resonant shock test equipment to better meet test specifications.

In the discussion after a presentation on resonant bar shock testing at IMAC-XL, it was mentioned that dynamic vibration absorbers (also called tuned mass dampers or tuned absorbers) can be applied to resonant shock test fixtures. For the resonant bar application, shock practitioners in some cases might desire to suppress or alter the bending modes of the resonant bar. The authors recognized that some existing experimental test items at Sandia National Laboratories could be fashioned into a dynamic vibration absorber and decided to perform initial work on a resonant plate. This decision allowed the team to use existing hardware and finite element models to get familiar with the application of dynamic absorbers as quickly as possible. A short investigation was conducted into the effects of such a dynamic absorber on a resonant plate bending mode.

Background

Resonant plate is a type of resonant fixture laboratory test that exploits the transverse bending mode of a plate, whose frequency is pre-determined by selecting appropriate plate dimensions. The test article is usually attached to the center of the plate surface, and the plate is struck with a projectile on the opposite side. The resulting pulse and ringdown event, measured with an accelerometer and analyzed via the Shock Response Spectrum (SRS), is very repeatable. To facilitate a wide range of shock event simulations, a test laboratory will modularize this technique, resulting in several resonant plates of different dimensions in the lab inventory, each designed to respond at different frequencies. To keep costs low, simple square plates of aluminum are most often procured and adapted for use [1].

David E. Soine · Adam J. Bouma · Forrest J. Arnold
Sandia National LaboratoriesAlbuquerque, NM 87185
e-mail: desoine@sandia.gov; abouma@sandia.gov; fjarnol@sandia.gov

Alexandra Karlicek et al. (eds.), *Dynamic Environments Testing, Vol. 7*, Conference Proceedings of the Society for Experimental Mechanics Series, https://doi.org/10.13052/9788743801528-7

Simple aluminum plates exhibit low damping, leading to long shock test ringdown times on the order of 150 msec or greater. To better simulate field environments, a shorter ringdown time is often desired. Damping devices such as "damping bars" and constrained layer approaches have been developed in response to this need [2]. In addition, the force pulse applied to the resonant plate by the projectile is modified by selection of the projectile mass, velocity, and "programmer material." This activity to optimize the force input to the resonant plate is critical to adequately excite the bending mode of the plate while limiting the response at higher test assembly natural frequencies. Aerospace component qualification test specifications are tailored to these test methods and are usually specified by a simple straight-line SRS profile and a ± 6 dB tolerance band. The typical SRS specification has a +12 dB/octave slope up to the "knee frequency," where it flattens to zero slope at a specified SRS acceleration level. In the test lab, a plate bending frequency that corresponds to the knee frequency is chosen to perform the shock test [3].

Though these methods have generally proven successful for single-axis testing, there is a desire to expand resonant fixture shock techniques to provide a multi-axis response with higher fidelity to the field environment. For a resonant fixture to provide a controlled shock response in more than one axis requires a test assembly designed to have fundamental responding modes with more sophisticated, multi-axis motion [4,5]. These test assemblies must be designed and optimized via computer modeling and simulation. Sometimes, one of several responding modes will drive an undesirable response at the test article. A dynamic vibration absorber could be a design feature useful in disrupting that undesirable modal response [6].

Testing and Analysis

In this investigation, the effect of a cantilevered-mass dynamic absorber on the bending mode of the resonant plate was observed. The cantilevered mass was made up of a cylindrical pedestal and square plate mass, which when assembled was known to have a fixed base natural frequency near 500 Hz and a weight of 9.4 pounds; see Figure 1. The resonant plate dimensions were 23.75 x 23.75 x 1.25 inches, with an approximate natural frequency of 500 Hz. The plate was suspended by ropes in front of an air gun, with two rods bolted to the plate to enable the plate to be leveled with cords when loaded with the dynamic absorber; see Figure 2 below. No damping bars were installed. To measure the shock event, an accelerometer was placed in the center of the plate, where a test article would ordinarily be installed for a uniaxial shock test. The cantilevered mass was bolted to the plate with precision spacers between the pedestal and the plate, since earlier investigations demonstrated that the precision spacers linearize the interface at the base of the pedestal and improve correlation with a finite element model. Setup testing was performed to select the projectile weight (35 pounds,) projectile velocity (17 feet/second,) and programmer material (two 1/4 inch pieces of felt) for the rest of the testing.

Fig. 1 Cylindrical pedestal and square plate.

Fig. 2 Bare resonant plate assembly and air gun.

When the center of the plate was struck with the projectile, the bending natural frequency of the unloaded resonant plate in the test configuration was 483 Hz, based on the peak of the acceleration Fourier transform (FFT) magnitude.

To develop the cantilevered mass into a tuned dynamic absorber, the cantilevered mass was attached to a drop table to approximate a fixed base boundary condition. The mass was then struck with a mallet, and acceleration was measured. The cantilevered bending natural frequency was 564 Hz in its original state. Additional masses in the form of steel screws, nuts, and spacers were added to each side of the cantilevered mass to tune the frequency to match the plate frequency of 483 Hz; see Figure 3. Acceleration response of the bare plate struck by a projectile and of the cantilevered mass dynamic absorber on a fixed base is compared in Figure 4 below, which also includes an illustration of the bare-plate bending mode shape estimated from a finite element model of the plate.

The cantilevered mass was initially installed on the resonant plate, 4.5 inches to the left and 4.5 inches down (i.e., Y = –4.5, Z = –4.5) from the plate center; see Figure 5. At that off-center location, based on the bending mode shape, a reasonable amount of plate rotation was expected to interact with the bending of the dynamic absorber. A shock test was performed, and the FFT and Shock Response Spectrum (SRS) of the acceleration in the center of the plate was recorded and is shown in Figure 6. Also shown is the response of the bare resonant plate for comparison. Based on the FFT response, it appeared

Fig. 3 Cantilevered mass, tuned to 483 Hz by adding spacers and bolts.

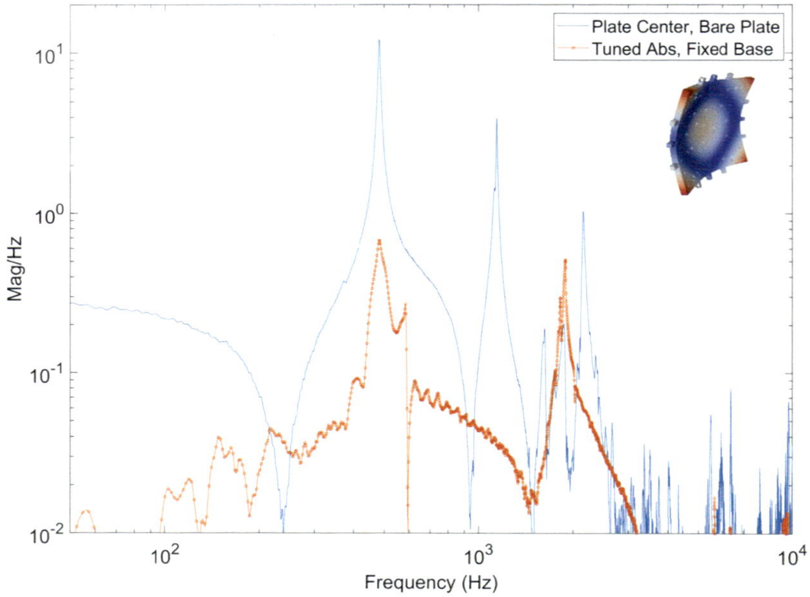

Fig. 4 Fixed base cantilevered mass (tuned absorber) frequency aligns with the resonant plate frequency at 483 Hz.

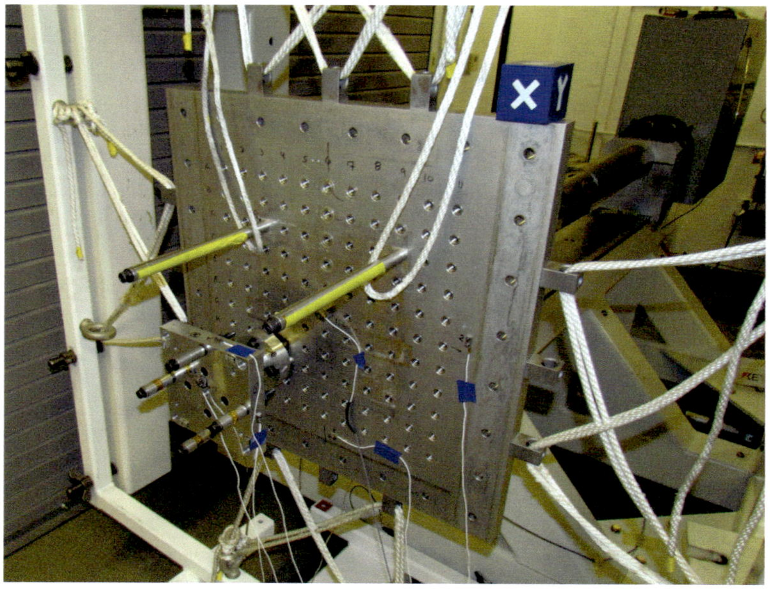

Fig. 5 Cantilevered mass offset from the center of the plate.

that the tuned absorber effect created two resonances in the vicinity of the original bending mode, at 357 and 557 Hz, with the 557 Hz mode dominating at approximately twice the level of the 357 Hz response. The relative amplitude of each peak is of interest when considering the SRS. The SRS of the assembly with the tuned absorber shows a "knee frequency" at 557 Hz, with a peak value 3.9 dB lower than that of the bare plate. Above approximately 400Hz overall SRS levels of the plate with dynamic absorber are about 2–3 dB lower than the bare-plate response.

To further investigate the cantilevered mass dynamic absorber assembly, the dynamic absorber was repositioned to (Y = –3, Z = –3) and (Y = –1.5, Z = –1.5). The resulting FFT and SRS of all four configurations is shown in Figure 7. It was expected that the dynamic absorber effect would be reduced as the absorber assembly was repositioned closer to the center of the plate, where the plate bending mode exhibits less rotation. The test results did not support those expectations, since the lowest knee frequency response amplitude occurred at when the dynamic absorber was at location (Y = –3, Z = –3).

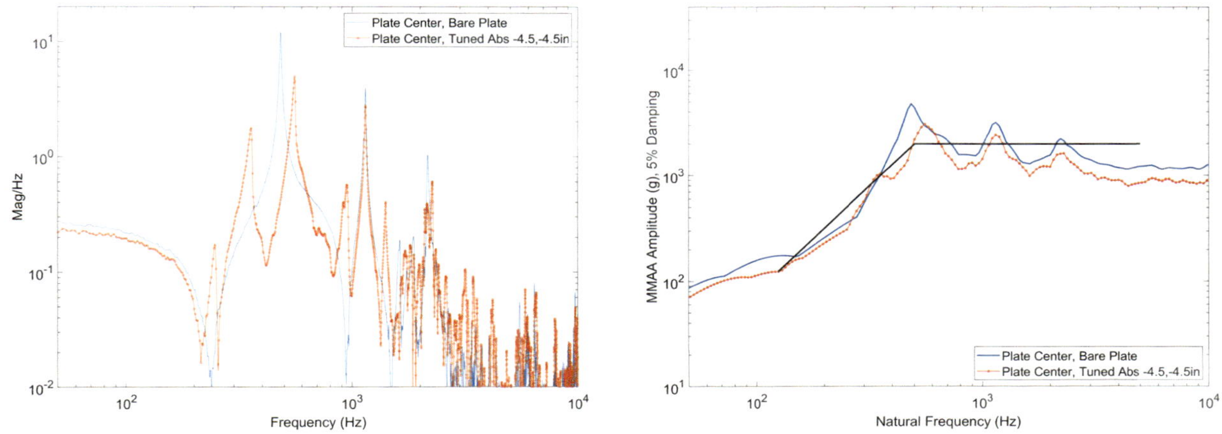

Fig. 6 Plate with tuned dynamic absorber compared to the bare-plate response.

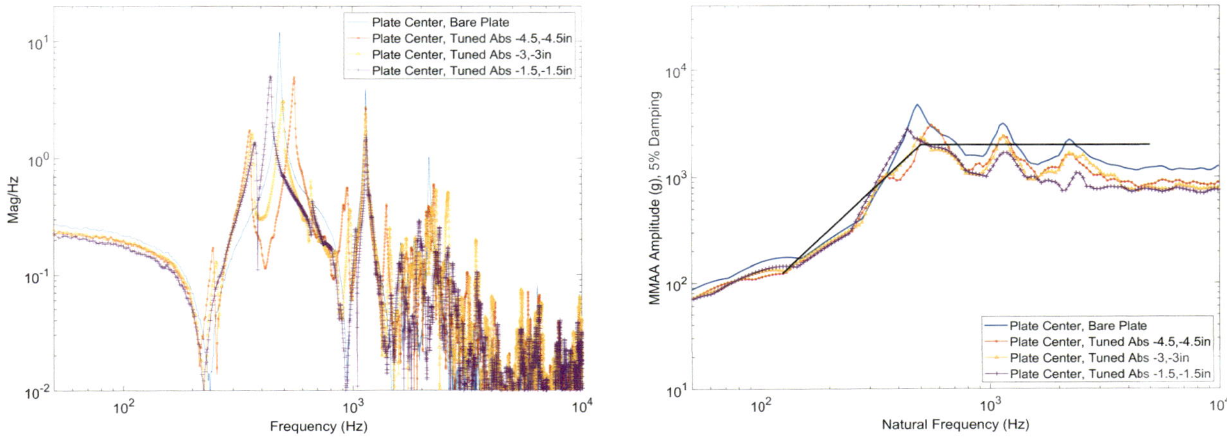

Fig. 7 Relocation of the tuned dynamic absorber on the plate.

Conclusion

The undamped cantilevered mass dynamic absorber modestly reduced the shock response when installed on an undamped resonant plate. In accordance with dynamic absorber theory, the assembly had natural frequencies above and below the target bare-plate mode when the dynamic absorber bending was well activated away from the center of the plate. The higher frequency mode responded with a greater amplitude than the lower frequency mode, and this effect was significant in the SRS analysis. As the dynamic absorber was moved closer to the center of the plate, the frequency of the dominant responding mode dropped and eventually was lower than the original bare-plate resonance, likely due to the absorber assembly mass moving closer to the original bare-plate bending antinode.

Acknowledgments This article has been authored by an employee of National Technology & Engineering Solutions of Sandia, LLC under Contract No. DE-NA0003525 with the U.S. Department of Energy (DOE). The employee owns all right, title and interest in and to the article and is solely responsible for its contents. The United States Government retains and the publisher, by accepting the article for publication, acknowledges that the United States Government retains a non-exclusive, paid-up, irrevocable, world-wide license to publish or reproduce the published form of this article or allow others to do so, for United States Government purposes. The DOE will provide public access to these results of federally sponsored research in accordance with the DOE Public Access Plan. https://www.energy.gov/downloads/doe-public-access-plan
This paper describes objective technical results and analysis. Any subjective views or opinions that might be expressed in the paper do not necessarily represent the views of the U.S. Department of Energy or the United States Government.

References

1. N. Davie, “The controlled response of resonating fixtures used to simulate pyroshock environments,” *The Shock and Vibration Bulletin*, Bulletin 56 Part 3, The Shock and Vibration Information Center, Naval Research Laboratory, August 1986.
2. M. Spletzer, and C. Sisemore, “Design of a resonant plate shock fixture to attenuate excessive high-frequency energy inputs,” *Proceedings of the 87th Shock and Vibration Symposium*, New Orleans, Louisiana, USA, 2016.
3. C. Sisemore, “Defining resonant plate shock test specifications in the time domain,” *Proceedings of the 90th Shock and Vibration Symposium*, Atlanta, Georgia, USA, 2019.
4. D. Soine, T. Schoenherr, J. Heister. “Evaluation of a Multiaxis Shock Fixture Concept.” *Dynamic Environments Testing, Volume 7, Conference Proceedings of the Society for Experimental Mechanics Series*, Springer, 2024.
5. D. Soine, T. Schoenherr, A. Bouma. “Resonant Bar Shock Test Equipment Under Offset Loading.” IMAC XLII, 2024.
6. Shock and Vibration Handbook, edited by C. Harris, 4th edition, McGraw Hill, 1996. Chapter 6.

Chapter 8
Analysis of a Tuned Vibration Absorber for Resonant Plate Shock Testing

Adam J. Bouma and David E. Soine

Abstract Vibration absorbers have vast applications in dynamic structures and can be applied to resonant fixtures for mechanical shock testing. Resonant structure techniques using plates or beams is an accepted method for simulating high-frequency pyroshock-type environments at the component and assembly level. In this paper, the resonant plate is the primary system, and a cantilevered fixture is used as a tuned vibration absorber to modify and tune the resonant frequencies of the plate. Model driven test design of the system is used to investigate the effect of the mass of the cantilevered fixture, mounting location on the plate, and pulse duration of the input force to modify the response of the resonant plate. Initial model validation and testing shows that the response of the resonant plate is modified with the addition of a tuned vibration absorber. The goal of this paper is to investigate the response of the system when using a vibration absorber for a mechanical shock test for the intent of tunability of the system.

Keywords Mechanical Shock · Tuned Absorber · Resonant Plate · Modal Analysis

Introduction

Dynamic vibration absorbers (DVAs) and Tuned Mass Dampers are mechanical attachments consisting of auxiliary masses with stiffness and damping to assist in controlling the amplitude of a vibrating system, termed the primary system [1]. The primary system can be protected from excessively high vibration levels with the use of tuned absorbers which can be included in the original system or added to an existing system. A DVA is also used to suppress vibration over a frequency band, generally in the vicinity of a targeted frequency. In this circumstance, a good trade-off between the suppressed original peak and the two newly emerged coupled peaks induced by the insertion of the absorber is crucial to obtain a global vibration reduction within the frequency band of interest [2].

The resonant plate shock test is a dynamic test where a test article is mounted on one side of the plate and a projectile is struck against the opposite side to simulate a mid-field pyroshock environment with the goal of exciting the breathing mode of the plate [3]. Using a tuned vibration absorber, the bending modes and breathing mode can be shifted in frequency or new modes can appear, depending on the application. The first three flexible modes of a free-free plate are shown in Figure 1 [4]. In this study, a cantilever type mass is used as a tuned vibration absorber to alter the response of the resonant plate and was used for previous studies on multi-axis response of mechanical shock tests [5]. Several parameters can be adjusted to

This article has been authored by an employee of National Technology & Engineering Solutions of Sandia, LLC under Contract No. DE-NA0003525 with the U.S. Department of Energy (DOE). The employee owns all right, title and interest in and to the article and is solely responsible for its contents. The United States Government retains and the publisher, by accepting the article for publication, acknowledges that the United States Government retains a non-exclusive, paid-up, irrevocable, world-wide license to publish or reproduce the published form of this article or allow others to do so, for United States Government purposes. The DOE will provide public access to these results of federally sponsored research in accordance with the DOE Public Access Plan https://www.energy.gov/downloads/doe-public-access-plan

Adam J. Bouma
Sandia National Laboratories. P.O. Box 5800 – MS0346Albuquerque, NM 87185
e-mail: abouma@sandia.gov

David E. Soine
Sandia National Laboratories P.O. Box 5800 – MS0557Albuquerque, NM 87185
e-mail: desoine@sandia.gov

Alexandra Karlicek et al. (eds.), *Dynamic Environments Testing, Vol. 7*, Conference Proceedings of the Society for Experimental Mechanics Series, https://doi.org/10.13052/9788743801528-8

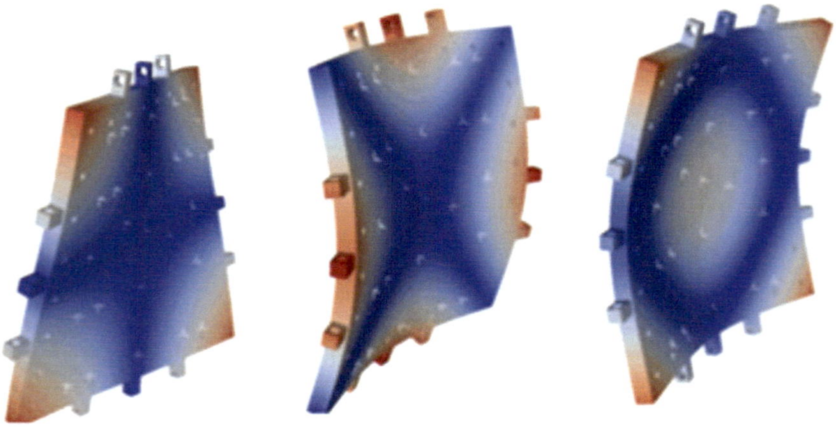

Fig. 1 First three modes of the resonant plate (twist, saddle, and breathing).

alter the modes of the plate, including tuning the mass of the cantilever fixture to the breathing mode of the plate, mounting location on the plate, and the pulse duration.

Tuned Vibration Absorber Design and Model Setup

For this analysis, the tuned vibration absorber used is a cantilever fixture consisting of a cylindrical adapter, surrogate fixture, and precision spacers with a weight of approximately 9.4 lb, shown in Figure 2a. The resonant plate used in this analysis is the 500 Hz plate, which is 23.73x23.73x1.25 inches in Figure 2b. Rope anchors are included, along with edge nodes representing the ropes that suspend the plate. Four point masses of 0.4125 lb each were added to each edge of the surrogate fixture to tune the fixture to the same frequency as the breathing mode of the resonant plate, shown in Figure 3. The untuned bending mode of the cantilever fixture is approximately 565 Hz. The breathing mode of the bare resonant plate and the bending mode of the tuned vibration absorber occurs at ~485 Hz. Several simulation runs were conducted, including impacting in the center of the plate with the fixture mounted in the center and also offset fixture configurations. The pulse duration of the input force will also affect the response of the system. The synthesized haversine pulse used for these simulations is shown in Figure 4, with a duration of 0.5 ms, 1.0 ms, 1.5 ms, 2.0 ms, and 2.5 ms and a magnitude of 10,000.

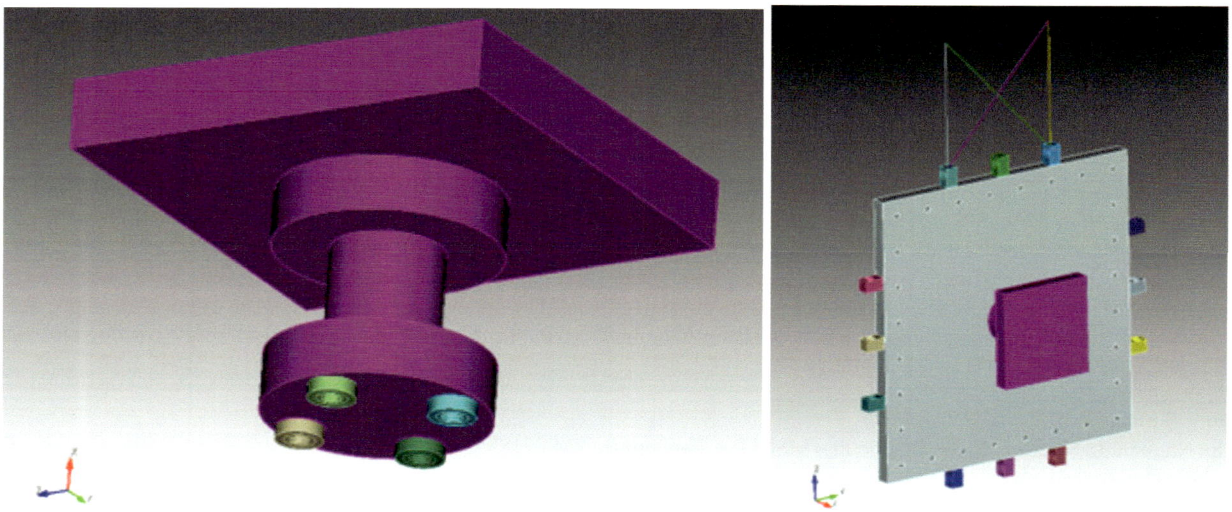

(a) (b)

Fig. 2 (a) Cantilever fixture with precision spacers and (b) cantilever fixture mounted in the center of the resonant plate.

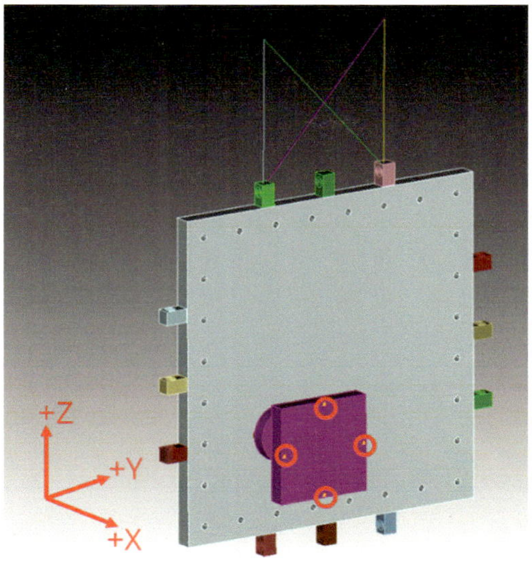

Fig. 3 Cantilever fixture mounted in the (-4.5, -4.5) configuration with point masses on each edge of the surrogate fixture.

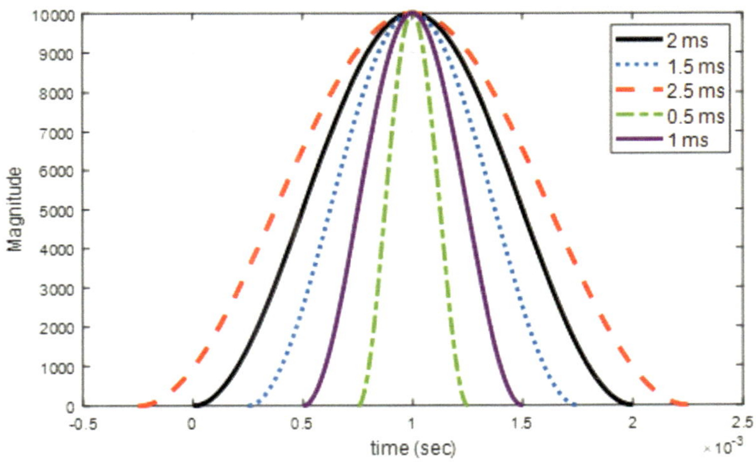

Fig. 4 Generated input force with varying pulse duration and 10,000 magnitude.

Simulation Results

The following results are based on comparisons made between the response of the resonant plate with a vibration absorber when changing the impact pulse and mounting locations on the resonant plate. By tuning the cantilever fixture to match the breathing mode of the resonant plate, the knee frequency is shifted to ~410 Hz in the Shock Response Spectra (SRS), shown in Figure 5. Tuning the fixture to match the breathing mode of the resonant plate shifts the response again, but only slightly, and the frequency peaks in the FFT is slightly higher than the SRS due to 5% damping. When moving the cantilever fixture to the offset configuration, two new resonance peaks appear at approximately 369 Hz and 571 Hz. This is seen in Figure 6, and the knee frequency is now driven by the 369 Hz response, and the magnitude is much lower than when the tuned absorber is mounted in the center of the plate. Figure 7 shows the response of the fixture when changing the pulse duration of the input force. A shorter duration pulse excites higher frequency modes while a longer duration pulse excites lower frequency modes, which is seen particularly in the x (axial) response of the SRS. The knee frequency will change depending on the input force.

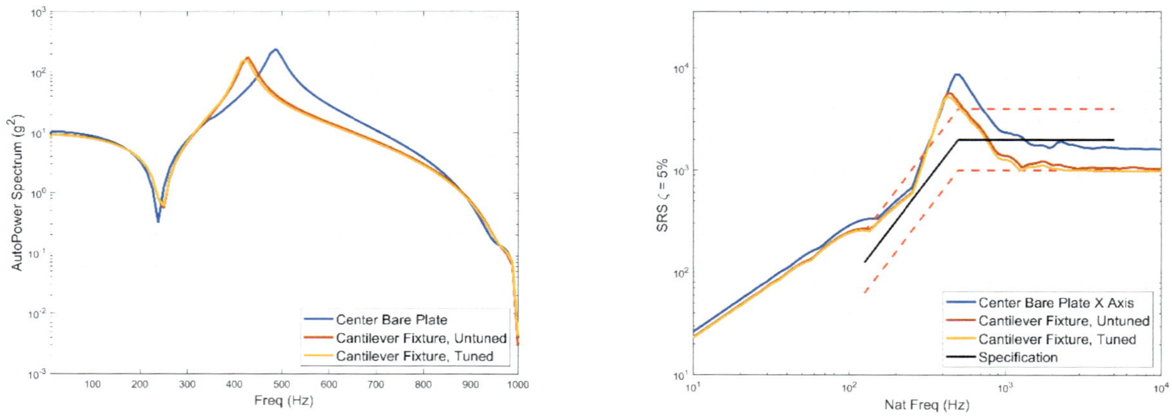

(a) (b)

Fig. 5 FFT and SRS of bare plate, untuned cantilever fixture, and tuned cantilever fixture with a center hit.

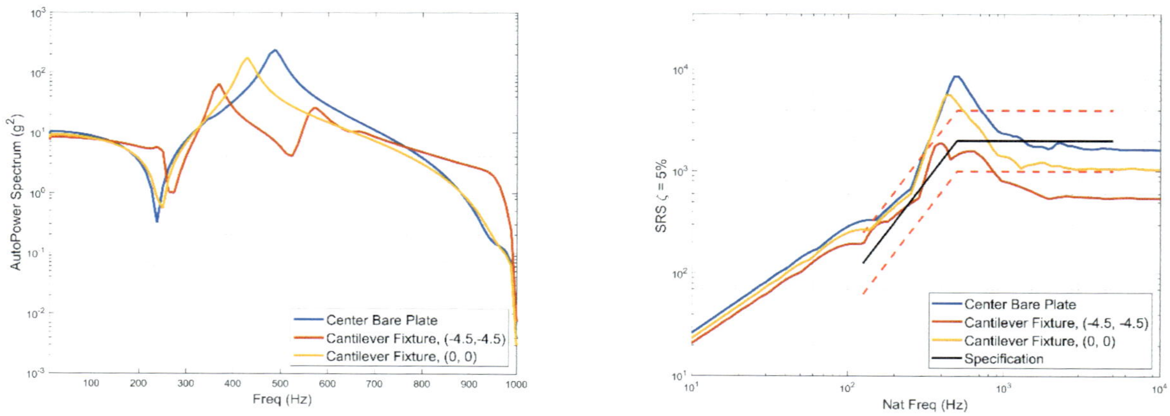

(a) (b)

Fig. 6 FFT and SRS of bare plate and cantilever fixture mounted in the center and offset configurations.

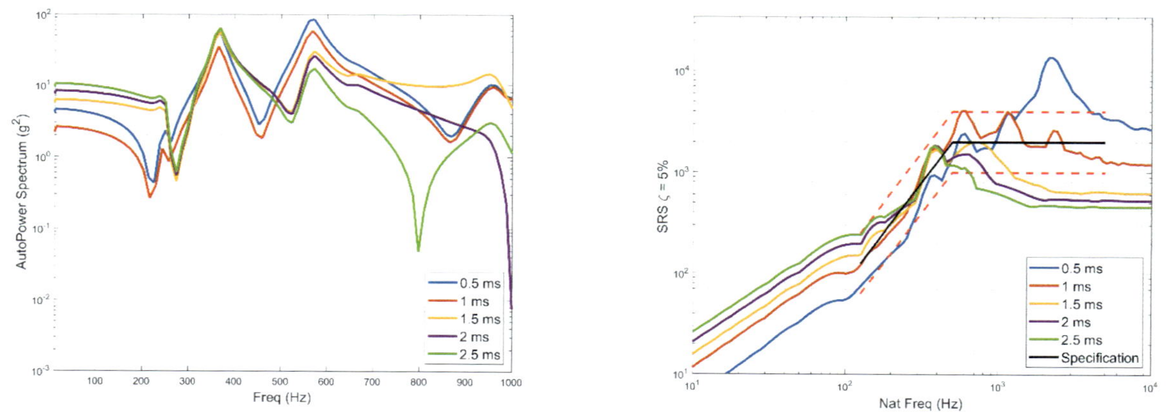

(a) (b)

Fig. 7 FFT and SRS of center hit with cantilever fixture mounted offset in configuration (-4.5, -4.5), comparing force input pulse duration.

Conclusions

The primary goal of this paper is to investigate the response of a resonant plate when adding a tuned vibration absorber. The bending mode of a cantilever fixture was tuned to the breathing mode of the resonant plate using point masses. The results presented show that the addition of the tuned vibration absorber will shift the knee frequency in the SRS, depending on the mounting location on the resonant plate. This is due to the tuned absorber creating two new resonant peaks, one lower and one higher than the breathing mode of the plate. Finally, the pulse duration will affect the response of the resonant plate and fixture. Further investigation should be pursued to optimize the design of a tuned vibration absorber for resonant plate shock simulation and testing.

References

1. Harris' Shock and Vibration Handbook, Cyril M. Harris, editor, Allan G. Piersol, editor. – 5th ed., 2002
2. Yang, C., Li, D., and Cheng, L., "Dynamic vibration absorbers for vibration control within a frequency band" Journal of Sound and Vibration, Volume 330, 2011.
3. Hopkins, R. and Sisemore, C., "Design of a resonant plate shock test for simultaneous multi-axis excitation" SAND2019-0777C, Sandia National Laboratories (SNL-NM), Albuquerque, NM (United States), 2019.
4. Schoenherr, T.F., Soine, D.E., and Witt, B.L., "Experimental modal analysis of a resonant plate during a mid-field pyroshock replication test" Sensors and Instrumentation, Aircraft/Aerospace Dynamic Environments Testing, Vol. 7, pp. 41–52, 2023.
5. Soine, D., Schoenherr, T., Heister, J., "Evaluation of a Multiaxis Shock Fixture Concept." Dynamic Environments Testing, Volume 7, Conference Proceedings of the Society for Experimental Mechanics Series, Springer, 2024.

Chapter 9
A Methodology for Feature Selection and Electrical Capability Prediction of a Coupled Shaker-DUT Model

D. Scheg, J. Heinlen, C. Garcia, L. Redmond, T. Roberts, A. Ramirez, and C. Haynes

Abstract In experimental pre-testing, it is often useful to predict the capability of an electrodynamic shaker to run an environmental test on a given device under test (DUT). In this work, a methodology is proposed for predicting shaker electrical capability through a coupled model and appropriate feature selection using Bayesian classification. The coupled model was formed through a reduced-order finite element model of a DUT and an experimentally-calibrated lumped parameter model of a shaker. A control problem was solved to calculate voltage spectral densities (VSDs) to meet the specified acceleration spectral density (ASD) for the test. Variance was given to model parameters and a Bayesian classifier was employed for feature selection. The Bayesian classifer was verified against three reduced-order model classes and the coupling methodology was validated experimentally. The paper culminates with an example problem which shows that the novel coupling methodology and the Bayesian classifier can be used to predict voltage requirements and select an unknown feature: joint stiffness.

Keywords Environmental testing · Bayesian classification · Reduced-order modeling · Shaker model · Finite element analysis

Introduction

In practice, it is common to run an environmental test on a device under test (DUT). An environmental test is the excitation of a DUT, usually by a shaker, under similar conditions that it would be subjected to in its true application. Often a test operator will not know the dynamic responses of the DUT (mode shapes, modal frequencies, etc.) or know if the shaker has the capability of performing the environmental test due to electrical constraints. Thus, it becomes advantageous to create analytical models for pre-test investigation. Past work investigated modeling a shaker as a lumped parameter system with electrical degrees of freedom (DOFs) to predict shaker requirements in a vibration test [1]. However, this investigation modeled the DUT as mass coupled to the shaker and did not consider the dynamic proper-

D. Scheg
Department of Mechanical Engineering, Purdue University, West Lafayette, IN 47907
e-mail: dscheg@purdue.edu

J. Heinlen
Mathematics & Computer Science Department, Whitworth University, Spokane, WA 992517
e-mail: jheinlen26@my.whitworth.edu

C. Garcia
Department of Mechanical Engineering, New Mexico Institute of Mining and Technology, Socorro, NM 87801
e-mail: chrisgarcia2003@gmail.com

L. Redmond
School of Civil, Environmental Engineering and Earth Sciences, School of Mechanical and Automotive Engineering, Clemson University, Clemson, SC 29634
e-mail: lmredmo@clemson.edu

T. Roberts · A. Ramirez · C. Haynes
Los Alamos National Laboratory, Los Alamos, NM 87545
e-mail: tproberts@lanl.gov; ajrami@lanl.gov; cmhaynes@lanl.gov

Alexandra Karlicek et al. (eds.), *Dynamic Environments Testing, Vol. 7*, Conference Proceedings of the Society for Experimental Mechanics Series, https://doi.org/10.13052/9788743801528-9

ties of the structure. A high fidelity finite element model (FEM) of the DUT can be used to capture these dynamics in the coupled model. However, a high fidelity FEM tends to have a large computational expense due to the high number of DOFs.

To reduce the computational cost associated with high fidelity models, various reduced-order models (ROM) may be implemented. There are many methods used for creating a ROM such as dynamic mode decomposition [2] and operator inference using deterministic linear regression [3]. In this paper, Guyan reduction (static condensation) [4] and component mode synthesis in the forms of Craig-Bampton and Craig-Chang [5] were used. All three of the used modeling processes begin with static condensation. Craig-Bampton and Craig-Chang also include generalized coordinates to represent a set of eigenmode amplitudes [4]. While both are accurate for static analysis, Craig-Bampton and Craig-Chang are more accurate at high frequency analysis, depending on the included eigenmodes.

ROMs, which retain a fraction of the original DOFs, introduce higher uncertainty, on top of the uncertainty present in the high-fidelity FEM. Specifically, joints have been notoriously difficult to model in the past because the global stiffness of the part is highly influenced by the flexibility of the joints. Past works have investigated the impact of these joints on the dynamics of a system [6–9].

This work proposes a novel methodology to select features, such as the most appropriate ROM or joint stiffness values, of a physically coupled shaker-DUT system, with the goal of determining the voltage needed to run an environmental test. To do so, a shaker model was coupled with a model of the chosen DUT, a box assembly with removable component (BARC). The shaker model was a lumped parameter model [1] and the BARC model was a ROM of a high-fidelity FEM. Given an acceleration specification, such as a benign acceleration spectral density (ASD) that operators are certain the shaker can complete, the voltage required to run the given test of the coupled system was calculated, in the form of voltage spectral density (VSD). The connection between the BARC base and removable component was modeled as sets of springs that were given varied stiffness coefficients. Then, using a Bayesian classifier, the joint stiffness model whose VSD most closely matched the experimental VSD was selected, showing the methodology proposed in this paper can be used to compare modeling techniques for a more accurate representation of a physical system.

Analytical Methods

As aforementioned, the purpose of this project was to: 1. understand the voltage requirements to run an environmental test using a coupled shaker-BARC model, and 2. select features of this system that best represent a set of physical test data. For this purpose, a high DOF model of the BARC system was previously created in Abaqus, a finite element analysis software. This work then implemented ROMs of the BARC to reduce computational expense. The equations of motion (EOMs) from the reduced BARC model and the shaker model were combined to represent the coupled system. The VSD of this coupled EOM was calculated, which estimates the voltage requirements to excite the BARC to the acceleration needed for an environmental test. In order to select features of this system, a Bayesian classifier was used. To demonstrate this method, the spring joint connections between the BARC base and the removable component were varied by using three stiffness coefficients, $k = 100, 1000$, and $10,000\ \mathrm{N/m}$. The Bayesian classifier selected the VSD from the class of stiffness coefficient that was closest to the experimental ground truth VSD.

Reduced-order modeling

The BARC was chosen the DUT because its dynamics are well known. The BARC was modeled in Abaqus using a fine mesh, consisting of over 100,000 DOFs. Three ROM techniques were implemented through Abaqus: Craig-Bampton, Craig-Chang, and Guyan reduction. These reduction methods require the user to specify which DOFs will be retained and fully calculated after reduction. Using engineering judgment, nine nodes were selected that were believed to capture the eight mode shapes of interest. A Modal Assurance Criterion (MAC) was performed, which concluded that the selected nodes adequately retained the overall dynamics of the structure, with an average MAC score of 0.9 for the eight modes. These retained nodes also reflected the positions on the physical BARC where accelerometers would be placed or where the shaker stinger would be connected (see Section 4.2). The first eight natural frequencies of each ROM were compared to the high-fidelity model, of which Craig-Chang had the lowest error. Consequently, this ROM was used for the demonstration of feature selection (joint stiffness) using Bayesian classification.

Joint modeling for classification problem

The joint model was chosen as a mechanism to demonstrate the feasibility of the Bayesian classifier to select appropriate features of a DUT. The joint model consisted of connecting the base of the BARC to the removable component using sets of springs with varying spring stiffness coefficients. Each connection to the base was modeled as four 3-DOF springs, 24-DOF in total. For simplicity of the initial demonstration, all 3 DOFs were assigned the same stiffness in each model class (e.g. the model with $k = 100\mathrm{N/m}$ assigns this stiffness in the x, y, and z direction of the springs).

Shaker model

To solve for the voltage required to excite a system to an acceleration specification, a lumped parameter shaker model was coupled to the FEM of the BARC. The shaker model consists of two mechanical DOFs: the armature and the stinger. It also contains one electrical DOF, the electromotive force, from the resistor and inductor (RL) circuit powering the mechanical components. Altogether, the system's equations of motion are

$$\begin{bmatrix} m_s & 0 & 0 \\ 0 & m_a & 0 \\ 0 & 0 & 0 \end{bmatrix} \begin{Bmatrix} \ddot{x}_s \\ \ddot{x}_a \\ \ddot{i} \end{Bmatrix} + \begin{bmatrix} c_1 & -c_1 & 0 \\ -c_1 & c_1 + c_2 & 0 \\ 0 & -K_f & L \end{bmatrix} \begin{Bmatrix} \dot{x}_s \\ \dot{x}_a \\ \dot{i} \end{Bmatrix} + \begin{bmatrix} k_1 & -k_1 & 0 \\ -k_1 & k_1 + k_2 & K_f \\ 0 & 0 & R \end{bmatrix} \begin{Bmatrix} x_s \\ x_a \\ i \end{Bmatrix} = \begin{Bmatrix} 0 \\ 0 \\ E \end{Bmatrix}. \tag{1}$$

See Table 1 for a complete list of parameters and their values. The electromotive force, E, was controlled to match a given acceleration specification. To calibrate this model, an electromechanical analysis of a the Modal Shop 2060E shaker took place.

These equations were solved using the direct frequency response function (FRF) method at each frequency line. The expression used for this method is

$$-w^2 \begin{bmatrix} m_s & 0 & 0 \\ 0 & m_a & 0 \\ 0 & 0 & 0 \end{bmatrix} + jw \begin{bmatrix} c_1 & -c_1 & 0 \\ -c_1 & c_1 + c_2 & 0 \\ 0 & -K_f & L \end{bmatrix} + \begin{bmatrix} k_1 & -k_1 & 0 \\ -k_1 & k_1 + k_2 & K_f \\ 0 & 0 & R \end{bmatrix} \tag{2}$$

[10],[11]. The mass of the stinger with impedance head attached, m_s, was directly measurable and the mass of the armature, m_a, was found through manufacturing specifications. All other parameters could not be directly measured so they were first estimated then calibrated. The estimation process stems from transfer function simplifications. The equations for these simplifications are as follows:

$$k_1 = (2 \cdot \pi \cdot f_h)^2 \cdot \frac{m_s \cdot m_a}{m_s + m_a}, \tag{3}$$

$$k_2 = (2 \cdot \pi \cdot f_l)^2 \cdot (m_s + m_a), \tag{4}$$

$$K_f = (m_s + m_a) \cdot A_m, \tag{5}$$

$$c_1 = \frac{k_1 \cdot K_f}{(2 \cdot \pi \cdot f_h \cdot (m_s + m_a) \cdot A_h)}, \tag{6}$$

$$c_2 = \frac{2 \cdot \pi \cdot f_l \cdot K_f}{A_l}, \tag{7}$$

where f denotes frequency (Hz) and A denotes the amplitude value at said frequency. The values for amplitude and frequency come from an acceleration over current FRF. The data for this FRF was established through experimental data where the shaker vertically excited the stinger and attached impedance head. Within these two parameters the subscript l denotes a value taken in the lower range of frequency, where the first mode shape can be observed. Similarly the subscript h is used for a value taken at a higher frequency where the second mode shape can be observed. Lastly, the subscript m is used for a value taken in the middle ranges of frequency where there are no modes present. Through analyzing the graphs in Figures 1a, 1b, and 1c, the lower frequency value chosen was 22Hz and the higher frequency value was 3977Hz. The middle frequency value was chosen at 220Hz, a relatively flat part of the FRF graph.

The values found using (3), (4), (5), (6), (7) produced a simplified representation of the system, but ultimately didn't reflect the dynamics of interest. For this reason, each parameter of the system, apart from known values m_s and m_a, was calibrated. This was done by first graphing the experimental data against the not optimized model FRFs: acceleration over current, acceleration over voltage, and impedance. Next, each parameter was optimized using particle swarm optimization or sequential quadratic programming. For each function, upper and lower frequency bounds were set to restrict the optimization to a specific range of the graph, as different frequencies in the FRF functions tended to be affected by the parameters. After the optimization was performed, the calibrated EOM was used as the shaker surrogate in the coupling process. Graphs of each FRF can be seen in Figures 1a, 1b, and 1c.

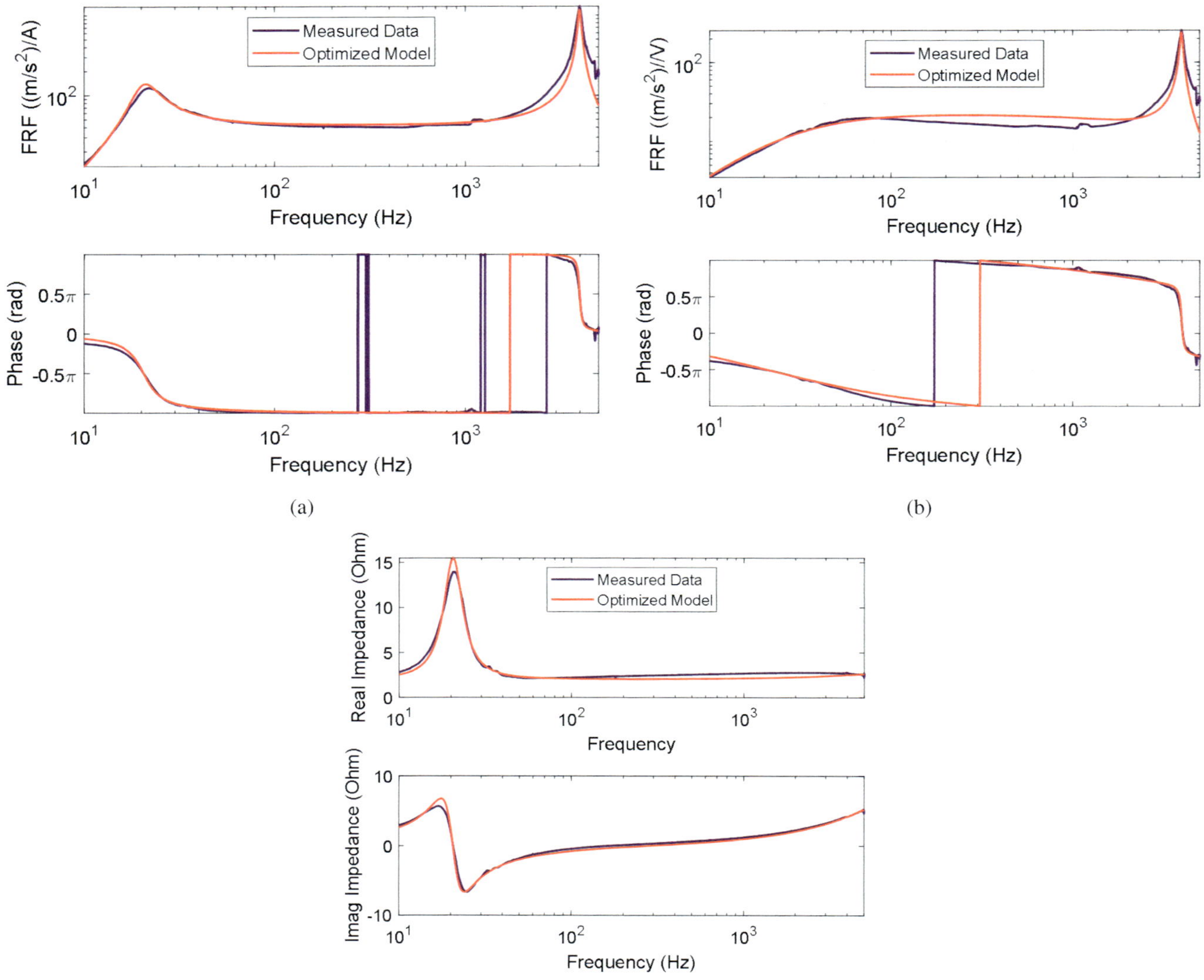

(c)

Fig. 1 Shaker model FRFs comparing the measured experimental data and the optimized model for (a) acceleration over current, (b) acceleration over voltage, and (c) impedance.

Coupled system model

In order to model the BARC and the shaker as one coupled system, their mass, damping, and stiffness matrices were first combined into a block diagonal matrix. The upper-left portion of each matrix was from the reduced BARC model and the bottom-right portion was from the calibrated lumped parameter shaker model (3x3). Additionally, in the global stiffness matrix, an arbitrarily high stiffness value, known as k_{glue}, was added at the DOF of excitation to represent the connection between the BARC and stinger. This is the point in the physical system where the shaker's stinger connects to the BARC. This value was also added to the stinger location on the shaker portion of the global matrix and symmetrically, k_{glue}, was subtracted from the off diagonal terms. While the full coupled matrix is to large to show, an example stiffness matrix of the BARC could be written as

$$\begin{bmatrix} k_{11} & k_{12} & k_{13} \\ k_{21} & k_{22} & k_{23} \\ k_{31} & k_{32} & k_{33} \end{bmatrix}. \tag{8}$$

Coupling the example matrix 8 with the stiffness matrix from equation 1 at DOF 3 gives the example coupled matrix of

$$\begin{bmatrix} k_{11} & k_{12} & k_{13} & 0 & 0 & 0 \\ k_{21} & k_{22} & k_{23} & 0 & 0 & 0 \\ k_{31} & k_{32} & k_{33}+k_{glue} & -k_{glue} & 0 & 0 \\ 0 & 0 & -k_{glue} & k_1+k_{glue} & -k_1 & 0 \\ 0 & 0 & 0 & -k_1 & k_1+k_2 & K_f \\ 0 & 0 & 0 & 0 & 0 & R \end{bmatrix}. \tag{9}$$

Because the overarching goal of this project was to model a shaker's ability to run an environmental test with a given DUT, a flat ASD of 0.0125 $(\mathrm{m/s^2})^2/\mathrm{Hz}$ was controlled at a specified point on the BARC, on both the physical and modeled BARC. The specified point was the retained node (see Section 3.1) on the removable component of the BARC. The FRFs from this data were calculated using methods described above (see Section 3.3), and were algebraically manipulated to solve for the electromotive force, resulting in VSD: power spectral density ($\mathrm{V^2/Hz}$) as a function of frequency (Hz). VSD is the form of data that was used to compare classes within the Bayesian classifier. Additionally, the coupled system model was validated against the experimental ground truth by comparing their VSDs and showing they qualitatively matched (see section 4 and Figure 5).

Bayesian classification methodology

Bayesian classification theory

A Naïve Bayesian classifier works by classifying an object X as class C when the posterior probability is highest, that is, when $P(C_i|X)$ for $i = 1, ..., m$ is maximum. Bayes' Theorem shows that

$$P(C_i|X) = \frac{P(X|C_i)P(C_i)}{P(X)}. \tag{10}$$

$P(X)$, the probability that X is the observation chosen, is constant in equation 10. All classes are assumed to be equally likely, that is, $P(C_1) = P(C_2) = ... = P(C_m)$ - a valid assumption since a constant number of observations was used from each class. Therefore, in equation 10, the only non-constant value to be maximized on the right-hand side is $P(X|C_i)$ [12].

In Naïve Bayesian Classification, it is assumed that all m points in each observation are independent of each other. The densities of each class's multivariate normal distribution were calculated under this assumption. These densities were normalized by dividing each class by the total density of all classes, resulting in the probability that the observation came from class C, or $P(X|C_i)$. The class with the maximum probability/density for the observation was selected as the best class.

Application of bayesian classification

As mentioned above, this project used Bayesian classification to select features of a shaker-DUT model that most closely matched the physical system, with the goal of determining if a shaker is adequate for an experimental test. As voltage was one of the limiting factors of a shaker test, VSDs were chosen as the medium to compare models. That is, the Bayesian classifier would select the class of model whose VSDs most closely matched the ground truth VSD. As an initial verification, coupled systems using three different ROMs of the BARC were compared to a sampled VSD coming from the VSD distribution of the shaker-DUT model using Craig-Chang for the ROM. Once the classifier was verified to correctly choose the Craig-Chang representation of the BARC, three joint model stiffnesses were compared to the experimental VSD.

As mentioned previously, the classifier needs to know the distribution of VSDs from each class of shaker-DUT model being compared in order to determine the probability the physical data comes from each model class's VSD distribution. There was uncertainty in the shaker parameters as well as the BARC model, so these parameters were each assigned a normal distribution according to the uncertainty of the parameter. In Table 1, there is a list of these varied parameters, as well as the mean and variance of the normal distribution they were selected from. Mean was determined by the measured/optimized values, and standard deviation was 10% of the mean value. For each shaker-DUT model class, Latin hypercube sampling was used to sample from all parameter distributions and calculate the VSD resulting from each sampled set of parameters. Note that the parameters were varied from the same distributions shown in Table 1 for every model class.

The resulting coupled system VSDs for each class maintained a multivariate normal distribution. The VSD distributions for each ROM are shown in Figure 2. It was also found that classifying with only the frequencies corresponding to the first eight minima of the VSD, which usually correspond to the natural frequencies, was sufficient for distinction in classification.

Table 1 Parameter mean and standard deviation in the shaker and BARC model. Mean values were determined by the results of parameter optimization or measurement. Standard deviations were set to be 10% of the mean value. Stinger mass, armature mass, and resistance were determined not to have a significant amount of uncertainty, so they were not varied.

Parameter (units)	Symbol	Mean	Standard Deviation
Stinger mass (kg)	m_s	0.0252	0
Armature mass (kg)	m_a	0.2700	0
Stinger damping coefficient (Ns/m)	c_1	21.1654	2.1165
Flexure damping coefficient (Ns/m)	c_2	11.9088	1.1909
Stinger stiffness (N/m)	k_1	$1.4392 \cdot 10^7$	$1.4392 \cdot 10^6$
Flexure stiffness (N/m)	k_2	$4.9129 \cdot 10^3$	491.2915
Coupling coefficient	K_f	12.6722	1.2672
Resistance (Ω)	R	2	0
Inductance, real (H)	$Re(L)$	$9.0000 \cdot 10^{-5}$	$9.0000 \cdot 10^{-6}$
Inductance, imaginary (H)	$Im(L)$	$-9.0000 \cdot 10^{-7}i$	$9.0000 \cdot 10^{-8}i$
Young's Modulus (Pa)	E	$69 \cdot 10^9$	$69 \cdot 10^8$

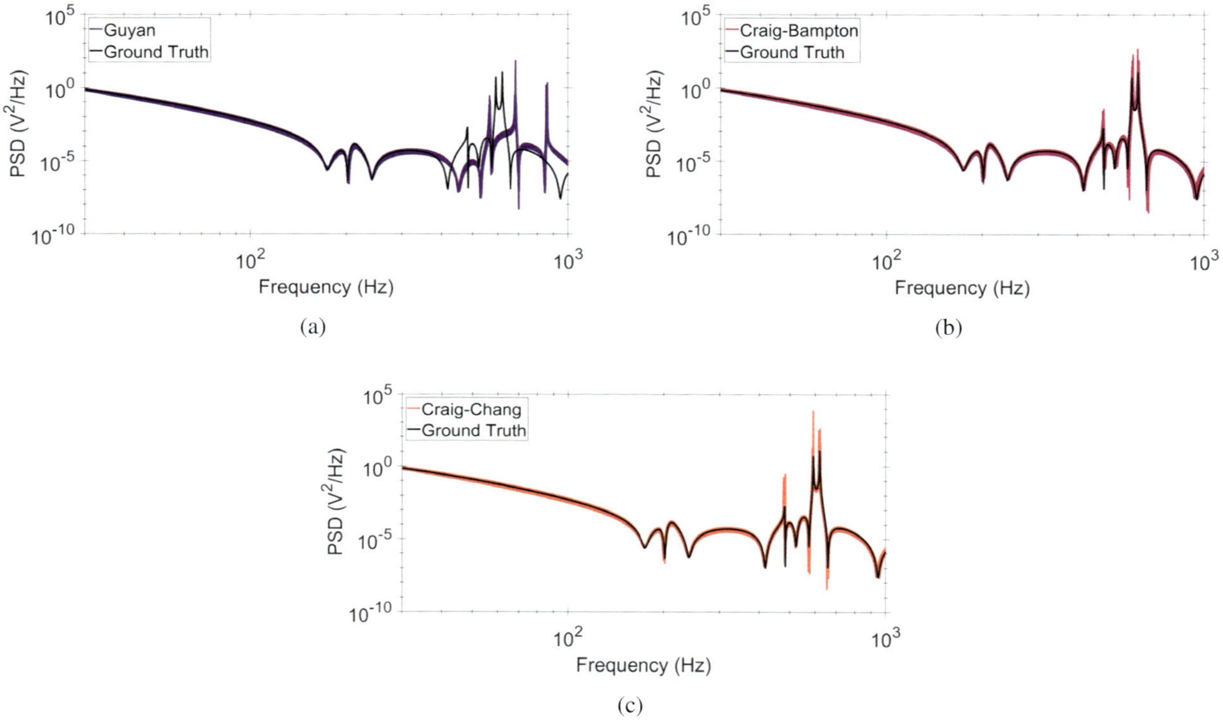

Fig. 2 VSD outputs of the three ROMs plotted against a Craig-Chang sampled ground truth. Ten outputs for each class are shown, however 75 were as (a) Guyan reduction (b) Craig-Bampton reduction (c) Craig-Chang reduction.

The magnitude of the VSD was not used for classification. Figure 3 is an example of points that were selected for classification in ten iterations of the VSDs generated using Guyan reduction. Note that although the 2-D graph is shown, only the frequency values were used for classification.

Initial verification of bayesian classification for coupled system selection

For initial verification, three reduced-order modeling methods were compared: Guyan reduction, Craig-Chang, and Craig-Bampton (see Section 3.1). For ground truth, a VSD from the Craig-Chang distribution was sampled (see Figure 2). The results of the Bayesian classifier can be seen in Table 2. The Craig-Chang ROM had, by far, the closest match to the sampled ground truth. This validated that the classifier could distinguish between VSDs of different classes and accurately make

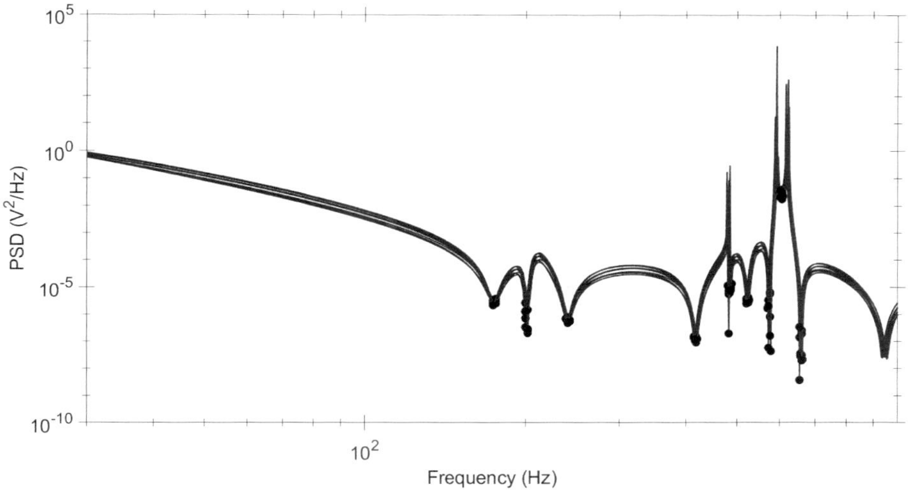

Fig. 3 A sample of VSDs from the normal distribution of VSDs calculated using Guyan reduction. The first eight local minima in the VSD were used for classification, and are highlighted as black dots.

predictions. After verification, the three joint stiffness values were compared and ground truth was determined experimentally, shown in Sections 4 and 5.

Table 2 Normalized and unnormalized probabilities output from the Bayesian classifier performed for verification on the three ROMs.

Class	Unnormalized Bayes' Prediction	Normalized Bayes' Prediction
Guyan Reduction	$e^{-24651.66}$	0
Craig-Bampton CMS	$e^{-39.13302}$	0
Craig-Chang CMS	$\boldsymbol{e^{-9.58068}}$	**1**

Experimental Methods

Shaker setup

Before coupling the shaker and BARC models (see Section 3.4), the shaker model was calibrated using experimental data. To do this, a stinger and impedance head were attached to the physical shaker and a broadband frequency from 5Hz to 6000Hz was provided, which was found to be adequate to detect both mode shapes. The first mode occurs when the armature and stinger are moving in phase, and the second mode occurs when the armature and stinger are moving out of phase. During the experimental test, voltage, current, and displacement were measured. Consequently, three FRFs were calculated as described in Section 3.3 and as seen in Figures 1a, 1b, and 1c.

BARC setup

To obtain the physical data used for the feature selection demonstration, the BARC was bolted to the shaker's stinger and was hung using fishing wire and rubber bands to imitate free-free boundary conditions (see Figure 4). The stinger of the shaker was bolted to the BARC at a point corresponding to a retained node in the ROMs (see Section 3.1). Similarly, accelerometers were attached at points also corresponding to retained nodes, including the chosen control point on the removable component of the BARC. The BARC was excited with a frequency range of 30 Hz to 1000 Hz, and Siemens Testlab was used to control to the specified ASD at the control point (see Section 3.4). Then, the VSD was computed and is used as the ground truth to compare against the VSDs of the modeled system.

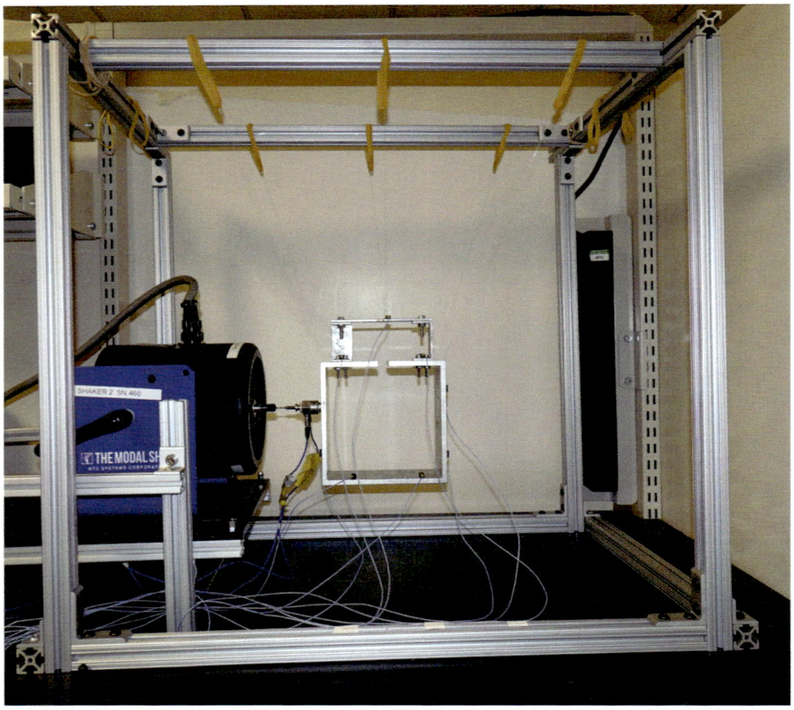

Fig. 4 The experimental set up with the BARC attached to the shaker in free-free boundary conditions.

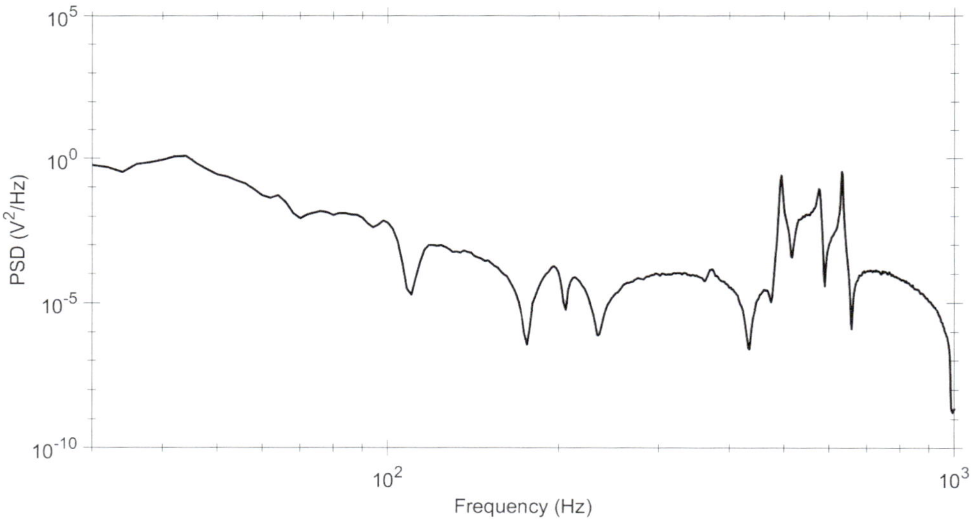

Fig. 5 The VSD from experimental data and used as ground truth in classification.

Analysis

After the coupled system was validated against the experimental ground truth and the Bayesian classifier was shown to distinguish VSDs of different ROMs (see Section 3.5.3), focus was turned to feature selection of the coupled model. Specifically, three values of joint stiffness between the base of the BARC and the removable component were compared: $k = 100, 1000$, and $10,000$ $(\mathrm{N/m})$ as shown in Section 3.2. 75 iterations were generated for each of the three classes, adding synthetic noise from a normal distribution to each iteration. The VSDs were then calculated and Figure 6 shows 10 representative iterations of each class plotted against the experimental ground truth.

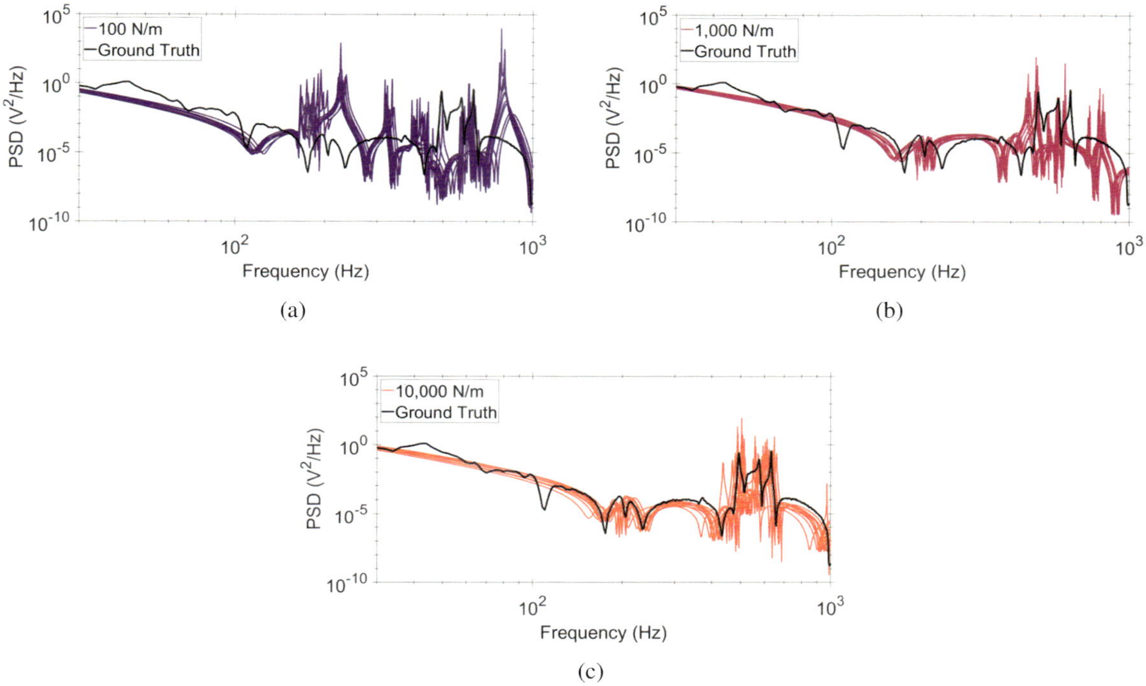

Fig. 6 VSD outputs of the three different spring stiffness coefficients plotted against the experimental ground truth. (a) Stiffness of 100 $\mathrm{N/m}$ (b) Stiffness of 1000 $\mathrm{N/m}$ (c) Stiffness of 10000 $\mathrm{N/m}$.

The model failed to pick up the resonance that is occurring at $\sim$140 Hz and some of the oscillatory behaviors of the voltage quantity in the experimental ground truth. This is most likely a byproduct of the simplifications that were made in the lumped parameter model of the shaker, the simplification of the joint stiffness as being identical in all 3 directions, and also uncertainty in the Abaqus model of the BARC. The uncertainty in the FEM model of the BARC comes from the reduced order methods, indicated by the MAC diagonal values not being correlated to a value of one. However, qualitatively, this verifies a few components of the model. Firstly, the two systems, the Abaqus mass and stiffness and the lumped parameter model, are being coupled correctly. Secondly, the lumped parameter model that was calibrated using experimental data is sufficient to pick up the dynamic responses of the experimental system. Lastly, at higher stiffness values of the joint connections the Abaqus model of the BARC is tracking the dynamic responses of the experimental setup adequately. Interestingly, when visually inspecting the system at the fundamental frequency in Abaqus with a 100 $\mathrm{N/m}$ stiffness value given for the springs, the top removable component separated from the box assembly base. This is most likely why there is less of a correlation to the ground truth in Figure 6a in comparison to the higher stiffness classes in Figures 6b and 6c.

These distributions were then imported into the Bayes' classifier and the minima were used to predict which of the three models best matched the ground truth. The classifier results shown in Table 3 predicted that the 10,000 $\mathrm{N/m}$ stiffness class most closely matched the experimental ground truth. The normalized probability of 1 shows that the joint connection of the physical system is best represented by a high stiffness value.

Table 3 Normalized and unnormalized probabilities output from the Bayesian classifier performed on three spring stiffness values.

Class	Unnormalized Bayes' Prediction	Normalized Bayes' Prediction
K value = 100	$e^{-488385.36}$	0
K value = 1,000	$e^{-3117.868}$	0
K value = 10,000	$\boldsymbol{e^{-488.1526}}$	**1**

Conclusion

Coupling a shaker model with a ROM of a DUT allows a user to asses the shaker's voltage capability to run an environmental test. Not only can this tool be used to inform a user's choice of shaker, but also to ensure that the structural dynamics of the DUT will be accurately measured during the environmental test. In addition, a Bayesian classifier can be used to select features, such as spring stiffness, of the DUT model that best represent the physical system. In a future work, this tool may be used to compare and select other model features, such as joint modeling techniques. Additionally, future work could explore limiting facets of a shaker other than voltage, such as force.

Acknowledgments We would like to gratefully acknowledge: Shashir Sharma for assisting with the Bayesian classifier, Stephen Wright for lending a Python to Abaqus wrapper, and Sean Kelly and Shannon Danforth for the help with reduced-order modeling.
This research was funded by Los Alamos National Laboratory (LANL) through the Engineering Institute's Los Alamos Dynamics Summer School. The Engineering Institute is a research and education collaboration between LANL and the University of California, San Diego's Jacobs School of Engineering. This collaboration seeks to promote multidisciplinary engineering research that develops and integrates advanced predictive modeling, novel sensing systems, and new developments in information technology to address LANL mission relevant problems.

References

1. Schultz, R. "Calibration of shaker electro-mechanical models". In Special Topics in Structural Dynamics & Experimental Techniques, Volume 5: Proceedings of the 38th IMAC, A Conference and Exposition on Structural Dynamics 2020, pages 133-144. Springer (2021)
2. Huhn, Q.A., Tano, M.E., Ragusa, J.C., and Choi, Y. "Parametric dynamic mode decomposition for reduced order modeling". Journal of Computational Physics, 475:111852 (2023)
3. McQuarrie, S.A., Guo, M., and Willcox, K. "Bayesian operator inference for data-driven reduced-order modeling". In 2022 Fall Western Sectional Meeting. AMS.
4. Flodén, O., Persson, K., and Sandberg, G. "Reduction methods for the dynamic analysis of substructure models of lightweight building structures". Computers & Structures, 138:49-61 (2014)
5. MORE, M. "Craig-bampton method" (2000)
6. Langer, P., Sepahvand, K., Guist, C., and Marburg, S. "Finite element modeling for structural dynamic analysis of bolted joints under uncertainty". Procedia engineering, 199:954-959 (2017)
7. Ma, X., Bergman, L., and Vakakis, A. "Identification of bolted joints through laser vibrometry". Journal of Sound and Vibration, 246(3):441-460 (2001)
8. Reid, J.D. and Hiser, N.R. "Detailed modeling of bolted joints with slippage". Finite elements in analysis and design, 41(6):547-562 (2005)
9. Ibrahim, R.A. and Pettit, C. "Uncertainties and dynamic problems of bolted joints and other fasteners". Journal of sound and Vibration, 279(3-5):857-936 (2005)
10. "Section 24: Frequency response analysis". https://help.autodesk.com/view/NINCAD/2022/ENU/?guid=GUID-FCA4E4B5-1A53-480E-A43A-A208E8F3C97E (2024) [Online; accessed 2024-09-25].
11. Kabe, A.M. and Sako, B.H. "Chapter 6 - experimental structural dynamics". In Kabe, A.M. and Sako, B.H., editors, Structural Dynamics Fundamentals and Advanced Applications, pages 429-511. Academic Press (2020)
12. Han, J., Kamber, M., and Pei, J. "8 - classification: Basic concepts". In Han, J., Kamber, M., and Pei, J., editors, Data Mining (Third Edition), The Morgan Kaufmann Series in Data Management Systems, pages 327-391. Morgan Kaufmann, Boston, third edition edition (2012)

Chapter 10
Motivating Multivariate Specifications for Multiple-Input, Multiple-Output Vibration Testing

Colin Haynes, Thomas Thompson, and Adam Watts

Abstract The concept of a conservative, bounding environment for multiple-input, multiple-output (MIMO) vibration testing is significantly more complex than in the traditional univariate case. In the univariate case, one-sided tolerance bounds provide a convenient definition of a conservative environment, expressing the level of conservatism in terms of coverage and confidence. Recent work has shown how the same concept cannot be applied directly to the multivariate case. However, the case of MIMO control is necessarily a multivariate one because the target responses are no longer considered independently. Rather than attempt to apply univariate methods to the MIMO case, this paper argues for adapting the definition of a conservative environment to better reflect the reality of the service environment.

Keywords Multivariate · Specifications · MIMO · Environmental Testing · Conservatism

Introduction

For many decades, electrodynamic shakers have been used to test aerospace hardware under simulated random vibration environments. Most such laboratory environmental testing is conducted on one axis at a time, though recently multi-axis methods have the subject of much study [1–4]. MIMO testing holds much promise for more accurately simulating the inherently multi-axial nature of most service environments. However, while established methods exist for developing specifications for single-axis testing, no consensus has been achieved yet regarding the appropriate methods for MIMO specifications. Other authors have noted the difficulty of determining MIMO specifications, but none to the authors' knowledge have considered the resulting multivariate nature of the statistics [5, 6]. Though there are likely other applications for the multivariate concepts discussed here, the primary focus of this paper is on the development of environmental specifications for MIMO structural dynamics testing.

First, the process for univariate specifications will be outlined briefly. Some multivariate statistical methods will then be examined to determine their suitability for deriving MIMO specifications. The concepts of coverage and confidence are revisited in a multivariate context, and several limitations to extending existing univariate approaches are discussed. Finally, a simple analytical example is used to demonstrate that the relationships between response variates is important to deriving an accurate environmental specification.

Univariate Specifications

Typically, aerospace structures are tested to environments that represent a statistically-conservative envelope of expected service environments to ensure a specified level of reliability in operation. An environmental specification for a given level of conservatism is most often determined by calculating the one-sided normal tolerance bound (or one-sided normal tolerance interval). The assumption for the mechanical testing application is that extreme low responses are not relevant for failure, and only extreme responses on the upper end of the distribution need to be considered in establishing a statistically-conservative specification. Mil-Std-810 (see method 514.8, Annex F and method 517.3) discusses the concept of the maximum predicted environment (MPE), which in that context is defined as a one-sided upper tolerance interval with coverage of 0.95 and

Colin Haynes · Thomas Thompson · Adam Watts
Los Alamos National Laboratory, Los Alamos, NM 87545
e-mail: cmhaynes@lanl.gov; tnthompson@lanl.gov; acwatts@lanl.gov

Alexandra Karlicek et al. (eds.), *Dynamic Environments Testing, Vol. 7*, Conference Proceedings of the Society for Experimental Mechanics Series, https://doi.org/10.13052/9788743801528-10

confidence of 0.50 [7]. Lalanne also elaborates on the process used to develop random vibration specifications [8]. Despite how common it is to define conservative levels in this way, it is still easy to misinterpret the meaning of such specifications [9].

Furthermore, this approach assumes that mechanical shock and vibration are defined as spectral quantities where each frequency line may be treated as its own independent distribution. Because single-axis testing is by far the most common experimental method, the specification development approach can assume a univariate distribution at each frequency line.

However, when moving to a MIMO test, it may be unwise to ignore the correlations between the several target responses (i.e., the spatial correlation). In general, for a distributed random vibration excitation of a structure, higher levels at one point would generally be correlated with higher responses at other points. However, this basic trend will be modified by the modes of the structure, as at certain frequencies some points on the structure will tend to higher responses than others based on mode shapes.

Figure 1 shows a standard normal distribution plot. Coverage in the univariate space is synonymous with the cumulative distribution function (CDF). Assuming a Gaussian distribution, the desired coverage is the area under the probability density function (PDF) curve. To establish that, one can perform and inverse CDF function on the desired coverage to obtain the data value that bounds the distribution. Total area under the PDF curve is equal to 1. Unfortunately, this is not the case in the multivariate space.

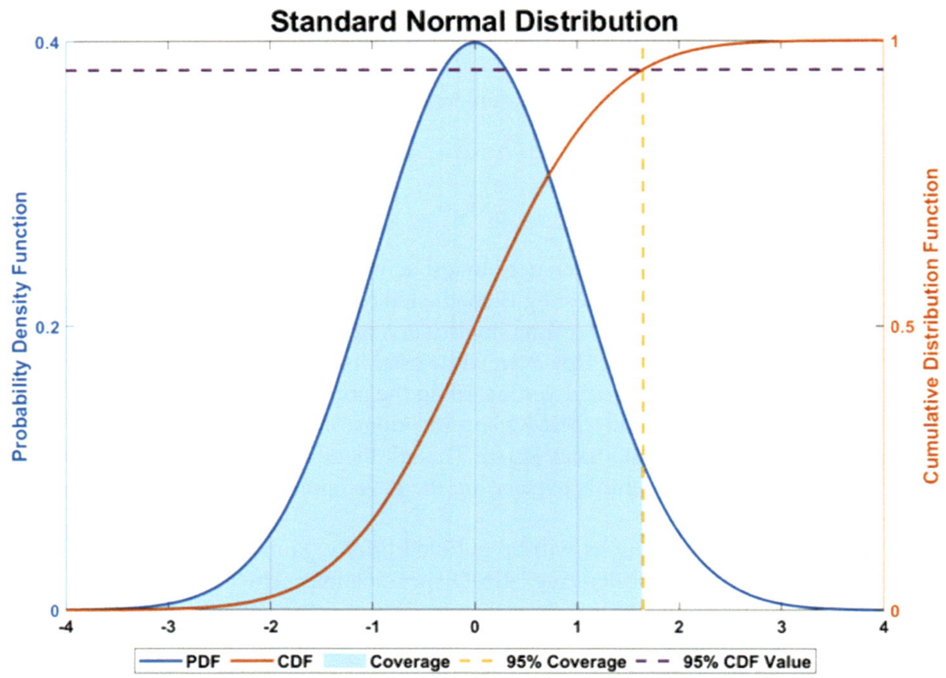

Fig. 1 Examples of PDF and CDF levels in the univariate case.

Simultaneous Prediction Intervals

Perhaps the clearest extension of the traditional univariate specification development approach to the multivariate case is the simultaneous prediction interval. In this case, the specification is based on meeting the defined coverage and confidence level on each variate independently, simply adjusting the confidence parameter by a Bonferroni correction that accounts for the number of variates and ensures that each variate is covered at the specified level [10]. While extremely efficient to compute, the Bonferroni correction does not consider any correlation between variates. Worse, the actual conservatism of the specification when considered relative to a known multivariate distribution increases as the number of variates increases.

Thus, although using more control DOF would generally improve match between a MIMO test and the field test reality, the simultaneous prediction interval would penalize using more control degrees of freedom by continuing to ratchet up the conservatism of the specification with every added variate. In this way, SISO tests would be specified at the lowest level of conservatism, despite yielding (in general) the poorest average match to the desired environment across the entire structure. We believe that a more sophisticated specification approach is required to take advantage of the potential of MIMO testing.

Dimensionality of Multivariate Specifications

Univariate tolerance bounds for a particular coverage proportion, usually denoted as β, are simply scalar. Multivariate analogies are necessarily of a higher dimension. Recall that one-sided tolerance bounds are equivalent to a one-sided confidence interval for a normal distribution quantile, denoted as $Q(\tau)$ (Meeker et al. 2023). Where $Q(\bullet)$ is the quantile function and τ is the quantile probability. Denoting the CDF of a random variable X, as $F_X(x)$, the quantile function for strictly monotonic distributions can be written as the inverse of the CDF, $Q(\tau) = F_X^{-1}(x)$.

For a univariate one-sided, upper, tolerance bound, has a particular case where $\beta = \tau$, i.e., the one-sided proportion of the population equals the quantile probability. The solutions for the intersection of the tolerance bounds and the quantile functions are one dimension less than the CDF for the respective distribution. For the normal univariate case, the CDF is a curve with the following mapping $F : \mathbb{R} \rightarrow [0,\ 1]$. Thus, the solution for the quantile function for a particular probability is just a scalar. However, the for the bivariate case, the CDF is a surface with the following mapping $F : \mathbb{R}^2 \rightarrow [0,\ 1]$. Denoting the joint CDF of two random variables X and Y, as $F_{XY}(x, y) = P(X \leq x,\ Y \leq y)$, this implies $Q(\tau) = F_{XY}^{-1}(x, y)$. The solution for the quantile function $Q(\tau)$ for a particular probability is now a curve that is comprised of infinite set of (x, y)vector tuples. Going up in dimensionality to a trivariate distribution yields a CDF that is a volume and the solution for the quantile function $Q(\tau)$ being a surface. For a given target multivariate quantile probability, τ, there exists an infinite set of target responses that satisfy the statistical requirements of being a solution to the quantile function. Methods do exist for determining the hypersurface and selecting reasonable points on that hypersurface that may be most relevant for environmental testing [11].

Furthermore, it should be noted that univariate tests always have a degree of tolerance associated with them to make the test achievable. Tolerancing guidelines for MIMO tests have not yet been established, but current state of the art MIMO practice also tends to allow the cross spectral density terms to vary to find a specification that can be achieved in the lab. Because multiple solutions can be described which satisfy the intent of the statistical conservatism, we suggest that it is best to specify the target environment as a space of solutions, allowing the particular solution to be chosen based on practicality of achieving the specification. Taking advantage of a more flexible specification may be challenging in implementation, but it is not necessary to find the optimal solution. Nominal specifications within the space of acceptable solutions could be prescribed if narrowing a range of choices is difficult *in situ*. Rather, being clear about what aspects of the target environment may be changed while still meeting the testing intent is a benefit in itself. If deviations from a nominal specification become necessary to execute a given test, outlining a space of solutions can provide the experimentalist with more options to perform the most appropriate test possible.

Coverage in a Multivariate Space

Figure 2 shows an example of a bivariate normal PDF and CDF. For simplicity, let us assume a bivariate gaussian distribution with a mean vector of 0 and identity covariance matrix. The PDF is a surface and, in this specific case, constant PDF values would be circles. If the covariance matrix has values in the off-diagonals or if the variances are not unity, the constant PDF values would be ellipses. A CDF curve at a constant value has two asymptotes (one parallel to the x-axis and the other parallel to the y-axis). Additionally, the volume under the PDF surface is equal to 1 (area under the curve for the univariate case and as the variates increase in dimensionality, so does this area/volume metric).

This is where we see a break between the univariate space (where the CDF value and coverage a synonymous) and the multivariate space. In the bivariate example, coverage would be akin to removing sections of the surface and then calculating the volume (which would be less than 1). Assume that a CDF contour is specified as our statistic of choice for designing a specification. In that case, calculating the coverage would involve removing the volume of the PDF curve that is above this CDF curve.

Figure 3 shows a CDF contour of 0.5 where the volume is removed from the PDF surface. The volume of this surface would be considered the coverage. To see the full effects of how the CDF affects coverage, a numerical study was conducted that varied the CDF values along with varying the correlation coefficient between variates 1 and 2.

Figure 4 shows the coverage (with the color map ranging from 0 to 1) as a function of the CDF value (y-axis) and the correlation coefficient (x-axis). The key observation of this plot is that holding a CDF value constant results in different

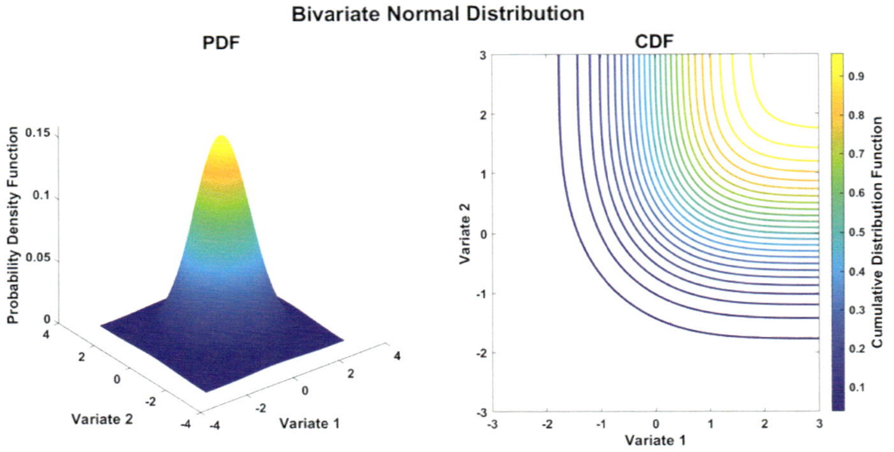

Fig. 2 Example PDF and CDF levels for a bivariate normal distribution.

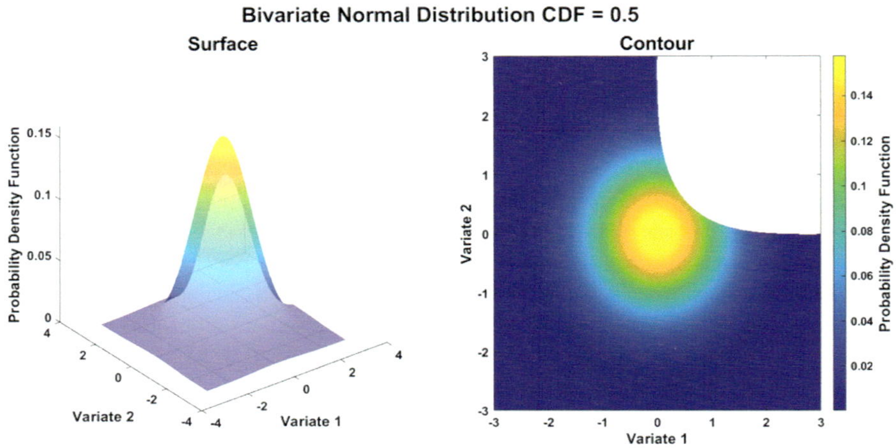

Fig. 3 Example PDF and CDF levels for a bivariate normal distribution, with volume removed at CDF=0.5.

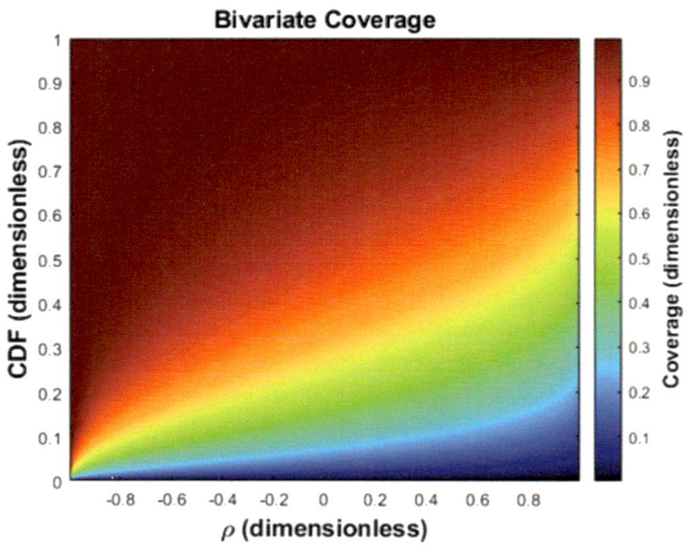

Fig. 4 Coverage as a function of CDF and correlation coefficient.

coverage values based on the correlation of the data. One does not simply select a coverage value in the multivariate space in the same way as one would in the univariate space.

While coverage in the multivariate space is not the same compared to the univariate space, analysts have some benefit to utilizing multivariate statistics. Specifications can either be based on a CDF contour or a constant coverage contour. Additionally, in the bivariate space, contours allow engineers a range of response points that satisfy a target level of conservatism, compared to a single response point in the univariate space. Having multiple points of the same statistical conservatism is important in case the ground test is not able to replicate some responses, but it is able to replicate others.

Numerical Demonstration

Synthesized Environments Data

For the purposes of this demonstration, consider a population of triaxial acceleration data from which a 3-variate target environment is derived. Realizations of the target environment observations consisted of zero-mean stationary random vibration 100 second time windows sampled at 8 kHz. Power spectral densities were then computed, and environments were derived. Because the power is constant with respect to frequency, one frequency point is considered for simplicity.

A correlation coefficient was selected based on experience with past test data. For X-Y, the coefficient was set to 0.3. For Y-Z, -0.3 was used. X and Z used a coefficient of 0, representing uncorrelated responses. Average PSD amplitudes were set to 0.0046, 0.0020, and 0.0005 G^2/Hz for X, Y, and Z, respectively.

To derive a MIMO environment from these synthetic data, a confidence value of 0.95 was first assumed. If a coverage of 0.9 is desired, simply computing that coverage level for each direction independently will not achieve the desired coverage for the multivariate environment. A comparison of the PSD levels determined computing based on univariate and multivariate statistics is given in Table 1. A multivariate coverage level of 0.7436 is included for comparison because that was computed to be the equivalent multivariate coverage level to using individual quantiles of 0.9 coverage. Note that "Multivariate" indicates the direct critical point method as defined by Watts et al., unless specified as the contour method from the same work [11]. Note that the derived environments levels and actual coverage of the distribution of possible acceleration responses are not equivalent for all methods.

Table 1 PSD levels computed using different statistical methods.

τ (Quantile)	Method	Critical point (G^2/Hz)		
		X	Y	Z
0.9	Univariate	5.973E-03	2.542E-03	6.339E-04
0.7436	Multivariate	7.967E-03	3.209E-03	8.047E-04
0.9000	Multivariate	8.974E-03	3.642E-03	9.162E-04
0.9000	Multivariate (Contour)	8.699E-03	3.425E-03	8.748E-04

Conclusion

This paper has described the some of the difficulties introduced in deriving specifications for MIMO, rather than single-axis, vibration tests. MIMO tests are more representative and accurate in part simply because they attempt to control more response DOF, so a simple multivariate generalization like the simultaneous prediction interval would be self-defeating. At the same time, these results have demonstrated that the correlation between variates (i.e., between different response channels in different locations and/or directions) has a significant impact on the resulting coverage level for a given CDF value. As a result, multivariate specification approaches are needed to take advantage of the full potential of MIMO vibration testing.

References

1. G. D'Elia, U. Musella, E. Mucchi, P. Guillaume, and B. Peeters, "Analyses of drives power reduction techniques for multi-axis random vibration control tests," Mechanical Systems and Signal Processing, vol. 135, 2020, doi:https://doi.org/10.1016/j.ymssp.2019.106395.

2. P. M. Daborn, C. Roberts, D. J. Ewins, and P. R. Ind, "Next-Generation Random Vibration Tests," in IMAC, A Conference and Exposition on Structural Dynamics, 2014, R. Allemang, Ed., 2014 2014: Springer International Publishing, in Conference Proceedings of the Society for Experimental Mechanics Series, pp. 397-410, doi:https://doi.org/10.1007/978-3-319-04774-4_37. [Online]. Available: https://link.springer.com/chapter/10.1007/978-3-319-04774-4_37
3. K. J. Moreno, V. V. N. Sriram Malladi, and P. A. Tarazaga, "A Data-Driven Approach to the Impedance Matched Multi-Axis Test Method," in Special Topics in Structural Dynamics & Experimental Techniques, Volume 5: Proceedings of the 38th IMAC, A Conference and Exposition on Structural Dynamics 2020, September 19, 2020 2021: Springer, Cham, pp. 335–339, doi:https://doi.org/10.1007/978-3-030-47709-7_31. [Online]. Available: https://link.springer.com/chapter/10.1007/978-3-030-47709-7_31
4. R. Mayes, L. Ankers, P. Daborn, T. Moulder, and P. Ind, "Optimization of Shaker Locations for Multiple Shaker Environmental Testing," (in en), Experimental Techniques, vol. 44, no. 3, pp. 283–297, 2020/06// 2020, doi:10.1007/s40799-019-00347-7.
5. G. Nelson, "A Systematic Evaluation of Test Specification Derivation Methods for Multi-axis Vibration Testing," in 36th IMAC, A Conference and Exposition on Structural Dynamics 2018, M. Mains and B. J. Dilworth, Eds., July 5, 2018 2019: Springer International Publishing, pp. 103–118, doi:https://doi.org/10.1007/978-3-319-74700-2_11. [Online]. Available: https://link.springer.com/chapter/10.1007/978-3-319-74700-2_11
6. R. Schultz and G. Nelson, "Techniques for Modifying MIMO Random Vibration Specifications," in Sensors and Instrumentation, Aircraft/Aerospace and Dynamic Environments Testing, Volume 7, Cham, C. Walber, M. Stefanski, and J. Harvie, Eds., 2023// 2023: Springer International Publishing, pp. 71–83.
7. Environmental Engineering Considerations and Laboratory Tests, MIL-STD-810H, 2019/01/31/ 2019. [Online]. Available: https://www.iest.org/Standards-RPs/MIL-STD-810H
8. C. Lalanne, Mechanical Vibration and Shock Analysis, Specification Development. John Wiley & Sons (in en), 2010, p. 517.
9. T. S. Edwards, "Probability of Future Observations Exceeding One-Sided, Normal, Upper Tolerance Limits," Journal of Spacecraft and Rockets, vol. 52, no. 2, pp. 622–625, 2015 2015, doi:10.2514/1.A33029.
10. K. Krishnamoorthy and T. Mathew, Statistical Tolerance Regions: Theory, Applications, and Computation, 1 ed. John Wiley & Sons, Ltd, 2009.
11. A. Watts, T. Thompson, and D. Harvey, "Confidence Intervals on Multivariate Normal Quantiles for Environmental Specification Development in Multi-axis Shock and Vibration Testing," ed: arXiv, 2024.

Chapter 11
Understanding Changes in Global Behavior Due to Control Location

Sarah Johnson, John Schultze, and Shannon Danforth

Abstract In distributed shaker, multi-axis testing, any number of locations and degrees of freedom can be used as control channels. Using differing control degrees of freedom can affect the global behavior of the test article, changing the response from the field. This study aims to examine how the response at monitored locations changes due to moving the controlled degrees of freedom on a test article. Initially, the control degrees of freedom are collocated with the drive point of a distributed shaker. Then, they are moved to a more compliant part of the test article, away from the actuator. To assess the quality of each test, the response at several monitored points will be compared to the field response using a multi-tiered approach. Understanding how control degrees of freedom affect the global behavior of the test article can help test engineers make more informed decisions about where to control and how to determine the quality of a test.

Keywords Multi-axis · Metrics · Aggregate ASD · MIMO · PSDPM

Introduction

Multi-axis vibration testing consists of exciting a structure in several directions simultaneously, as opposed to traditional single-axis testing which involves sequential tests in each principal axis of the test article. Multi-axis loading can be more representative of in-service environments because it allows greater flexibility in approximating boundary conditions. It also better approximates complex loading which can make the test environment more similar to the field environment. There are several methods to perform multi-axis vibration testing, and several methods for analyzing multi-axis vibration data.

Distributed shaker methods such as impedance matched multi-axis testing (IMMAT) involve attaching several shakers across the body of the test article and imparting force via a stinger [1]. Distributed shaker methods have largely been considered for flight environments where the test article experiences distributed loading across its surface. Previous researchers have had success matching flight relevant vibration levels using many control locations [2]. However, when any location on the test article can be a candidate shaker location, the design space of the test becomes quite large. It is also difficult to impart large forces and displacements using distributed shakers. There is no standard to define what is "good enough" in these styles of testing, primarily due to a lack of data from field testing. In addition, it can be difficult to assess the large amounts of data from these tests and determine how much better a multi-axis test approximated the field condition, especially when it comes to monitored locations that were not used for control.

In addition to distributed shaker methods, base-excited multi-axis vibration testing methods have also gained popularity. In a base-excited vibration test, the test article is rigidly attached to some sort of platform with a fixture, and the platform is excited in such a way that replicates the field environment. Bolting the test article to a shaker table with a rigid fixture, much like traditional single-axis testing techniques, reduces the complexity of the test set up and design. Base-excited methods can be used for transportation environments, where test articles are fixed to a platform and experience loading from the connection surface, a boundary condition that can be more readily replicated in a lab setting. Researchers have had success with commercial 6 degree-of-freedom shaker tables, custom-designed shaker platforms that have been used for component level testing, and and custom-designed shaker tables [3] [4].

Assessment of multi-axis testing can also be challenging. Due to the large number of control degrees of freedom and monitored degrees of freedom, it can be difficult to resolve what is a successful test. Kramer et al. used several metrics to try and assess how well multi-axis testing compares to single-axis testing [5], but the results are often limited to comparing one or

Sarah Johnson · John Schultze · Shannon Danforth
Los Alamos National Laboratory, Los Alamos, NM 87545
e-mail: sarahmjohnson2018@gmail.com; schultze@lanl.gov; sdanforth@lanl.gov

Alexandra Karlicek et al. (eds.), *Dynamic Environments Testing, Vol. 7*, Conference Proceedings of the Society for Experimental Mechanics Series, https://doi.org/10.13052/9788743801528-11

two control degrees of freedom at a time. Rohe et al. used an aggregate technique that adds all controlled autopower spectral density (ASD) responses and compares them to the sum of all control specifications [2]. Adding ASDs together results in much less data to compare but loses the information about how each channel responds. Also, there is no exploration of how the loss in resolution might affect the overall assessment of the test.

There are many difficulties with multi-axis testing, from implementation to assessment. This study aims to explore how changing the control location on a test article changes the response at monitored locations on the test article while utilizing a multi-axis test approach. In addition, this study will focus on data presentation can affect a test engineer's ability to quickly and accurately determine if a test was successful. We will assess the G_{RMS}, dB error (e_{dB}), and the frequency response function similarity metric (FRFSM) on aggregate response as well as individual response to determine what "level" of fidelity is needed in analysis to assess the global behavior of the system compared to what it sees in the field.

Testing

Field test

In this study, the Generic OBject for Laboratory Environmental Testing (GOBLET) is used as the test article. The GOBLET is an aluminum, three-piece assembly that is used to assess complex dynamics of cylindrical components. The GOBLET's three parts are an adapter plate, a cylindrical mass, and a bell, which are shown in Figure 1. The bell is thin-walled and is relatively compliant, especially when attached to the more rigid cylinder. The bell bolts onto the cylinder at 3 legs. The GOBLET adapter plate and cylinder primarily exhibit rigid body motion, with their axial motion being nearly identical to the input at the base of the assembly.

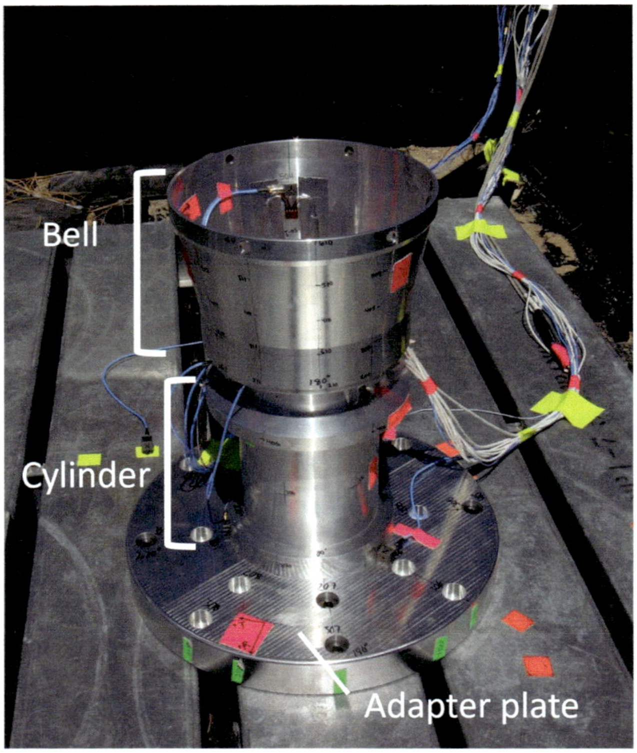

Fig. 1 Configuration of GOBLET during field testing.

To have a set of reference data, we field tested the GOBLET in a transportation environment. To perform this testing, six tri-axial accelerometers and three uni-axial accelerometers were attached across the GOBLET assembly. The bell piece had four tri-axial accelerometers, with two on one of the legs and two at different heights along the bell. The assembly was bolted to a metal pallet that secured in the bed of a truck, as shown in Figure 1. We then drove the truck around several paved and dirt roads to record different types of excitation at many levels.

To use the data from the field as an environmental test specification, we examined the time data and found a segment that had high levels of random vibration and did not contain shocks. Then, we computed a coarse ASD of the segment and used it as a comparison state between the different environmental tests. At frequencies above 500 Hz, the ASDs exhibited energy levels that were below the noise floor of the accelerometers used, so the high frequency data was excluded from analysis. Below 30 Hz, the ASDs exhibited displacement levels that were too high for the modal shaker used in this study. When creating specifications, we considered a truncated frequency range of 30 to 515 Hz.

Environmental test

To recreate the field condition in the lab, we used a technique involving a large, single-axis shaker and a distributed, modal shaker. The base excitation was provided by a Unholz-Dickie (UD) H560 single-axis shaker on which the test assembly was bolted. The placement and securing of the GOBLET on the single-axis shaker mirrors the boundary condition that the GOBLET experienced in the field closely. In addition to the base excitation, we also placed a distributed shaker, from here forward referred to as the *supplemental shaker*, on the test article. The location for the supplemental shaker was chosen based on lessons learned from prior testing, namely that the stinger drive block was most likely to stay attached to the cylinder due to its rigidity as opposed to the bell. The supplemental shaker was a 100-pound-force shaker from The Modal Shop with model number 2100E11. Excitation was imparted by a metal stinger which threaded into a custom drive block that was secured to the assembly using dental cement. The test set up is shown in Figure 2. Further information about this testing technique can be found in [6].

During the laboratory testing, two degrees of freedom (DOFs) were used as control channels. In the first set of tests, the control DOFs were located at the supplemental shaker drive point (see Figure 3(a)). The tests were controlled axially and radially by the UD shaker and the supplemental shaker, respectively. In the second set of test runs, the control DOFs were placed at a location 90 degrees from the drive location on the bell which is a more compliant region of the GOBLET

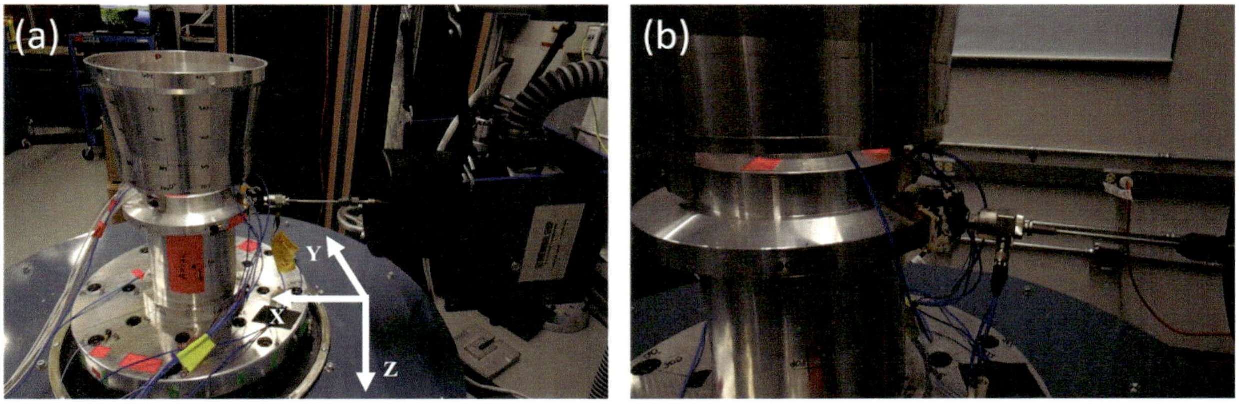

Fig. 2 GOBLET during laboratory testing with (a) full configuration and coordinate system and (b) drive block attachment.

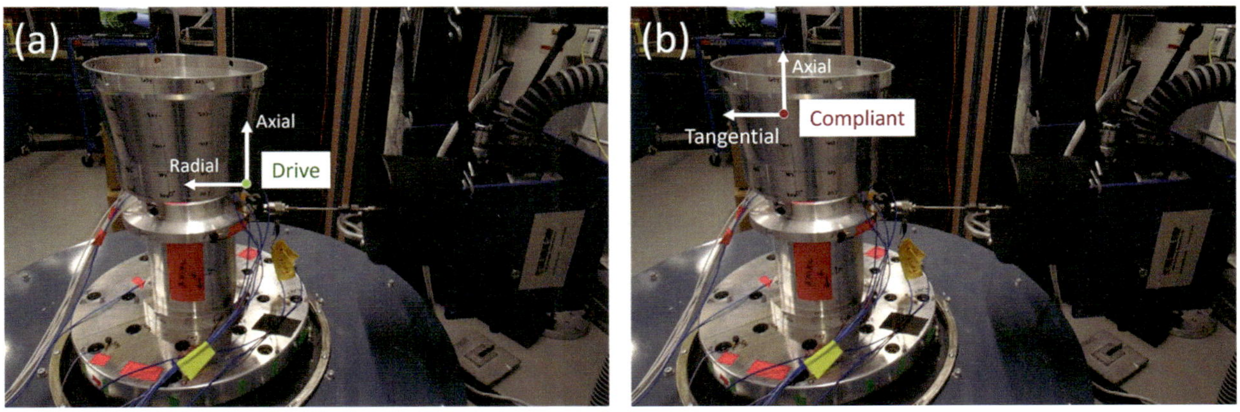

Fig. 3 Control degrees of freedom for (a) drive point control and (b) control from compliant location.

(see Figure 3(b)). One DOF was excited axially by the UD shaker, and the other DOF was excited tangentially by the supplemental shaker. It is important to note that the supplemental shaker was in the same location for both sets of tests.

Analysis

The goal of this analysis is to investigate what level of analysis allows us to detect which set of control DOFs results in the "best" test. For this case study, the best control location of the two will be determined by applying a series of metrics and assessing which is closest to the field response (explained further in next section). At the first level, we will take the sum of the ASDs of all channels and make comparisons between the field test and the global behavior based on the aggregate ASD. To compute the aggregate ASD, we will use the sum of autopower spectral densities used in [2] which will provide a single ASD curve consisting of the sum of the 21 channels present in the test. There will be one curve for the field response, one from the drive environmental test, and one from the compliant environmental test. When calculating the average ASD, we will only consider the DOFs that were present in both the field test and the laboratory test. The sensors corresponding to these DOFs were not moved or modified from the field configuration.

At the second level, we will assess the ASDs by direction. We will look at the sum of all X-direction ASDs, then Y-direction ASDs, and then Z-direction ASDs. Finally, at the third level, we will examine individual monitored location ASDs. For the sake of brevity, we will only examine the responses from one tri-axial accelerometer in this paper.

Our motivation for the tiered approach is to determine if a coarse analysis can provide a similar assessment of test performance when compared to a more detailed approach. The aggregate ASD analysis is a common strategy when test engineers need a quick assessment of a test, especially in situ. Through this study, we can improve confidence in this assessment technique as well as assess how different error metrics can affect the perceived performance.

Error metrics

One of the most common ways that vibration tests are assessed is using the G_{RMS}. The G_{RMS} of a autopower spectral density (ASD) curve represents the average energy across a frequency range and corresponds to the area under the ASD curve. The G_{RMS} is computed as

$$G_{RMS} = \frac{1}{n_{freq}} \sqrt{\sum_{i=1}^{n_{freq}} G_{xx}(f_i)^2}, \tag{1}$$

where $G_{xx}(f_i)$ is the auto spectral density value at the ith frequency line, for $i \in \{1, 2, ..., n_{freq}\}$. The G_{RMS} is often used because it simplifies the ASD to a single value, thus making test results easier to understand. However, it is sensitive to large amplitude response. This sensitivity can cause high amplitude excursions to appear much more severe than low amplitude excursions or under tests, resulting in an unreliable value.

Another common assessment of how well a test controlled is the decibel (dB) error. The dB error involves determining the dB difference between a response and some sort of reference. The mean dB error is given by

$$e_{dB}(f_i) = \frac{1}{n_{freq}} \sum_{i=1}^{n_{freq}} 10 \log \left(\frac{G_{xx}(f_i)}{G_{ref}(f_i)} \right), \tag{2}$$

where G_{xx} is the response ASD at the ith frequency line and G_{ref} is the specification ASD at the ith frequency line. Because the dB calculation involves a ratio of response to reference, dB error is equally sensitive to large amplitude excursions as it is to low amplitude excursions. We are interested in the total energy in the system for comparison, so we will be using the dB error instead of the RMS dB error or absolute dB error, which is another common test assessment.

Finally, we use a metric called the frequency response function similarity metric (FRFSM). While this method was originally developed to assess the differences in FRFs between analytical models and experiments by [7], the FRFSM can be used to compare any two signals at its basic level. Because we will be comparing ASD signals instead of FRFs, we will call this metric the power spectral density probability metric (PSDPM) for clarity. At each frequency line i (for $i \in \{1, 2, ..., n_{freq}\}$), a probability density function (PDF) $g(x_i, \mu_i, \sigma_i^2)$ is created:

$$g(x_i, \mu_i, \sigma_i^2) = \left(\frac{1}{\sigma_i \sqrt{2\pi}} \right) exp \left(-\frac{1}{2} \left(\frac{x_i - \mu_i}{\sigma_i} \right)^2 \right). \tag{3}$$

where the PDF is centered at a mean value μ_i with distribution width σ_i, and an error metric value of x_i. We will be using the dB error described in Equation 2 as the x_i value. The mean value of the PDF is set to zero, indicating perfect agreement between the response and specification, and the distribution is set such that 3σ on the PDF is equivalent to 6 dB, which is a common abort limit in environmental testing. The PSDPM value is then normalized by a PDF with a mean and error metric value of 0 to ensure the final value will be between 0 and 1:

$$\text{PSDPM} = \frac{1}{n_{freq}} \sum_{i=1}^{n_{freq}} \frac{g(e_{dBi}, 0, \sigma_i^2)}{g(0, 0, \sigma_i^2)} \quad (4)$$

where we will be computing the PSDPM at each frequency line then using the RMS value for assessment. The PSDPM offers a different perspective into the error between field and environmental response. At each frequency line, the PSDPM gives the probability that the dB error between the field and environmental response is zero, that is that the responses are the same. For comparison to other scalar-value metrics in this paper, we compute the RMS of the PSDPM across frequency lines. We consider this PSDPM RMS as an alternative measure of how well the specification/field environment was achieved by a given test.

Difference in response

Before we can make an assessment of how similar the global response of the GOBLET was between the lab and the field, we must first ensure that the control DOFs were controlled well in the environmental test. To assess this, we will first use visual inspection, then apply the error metrics from the previous section. In Figure 4, we see agreement when controlling to

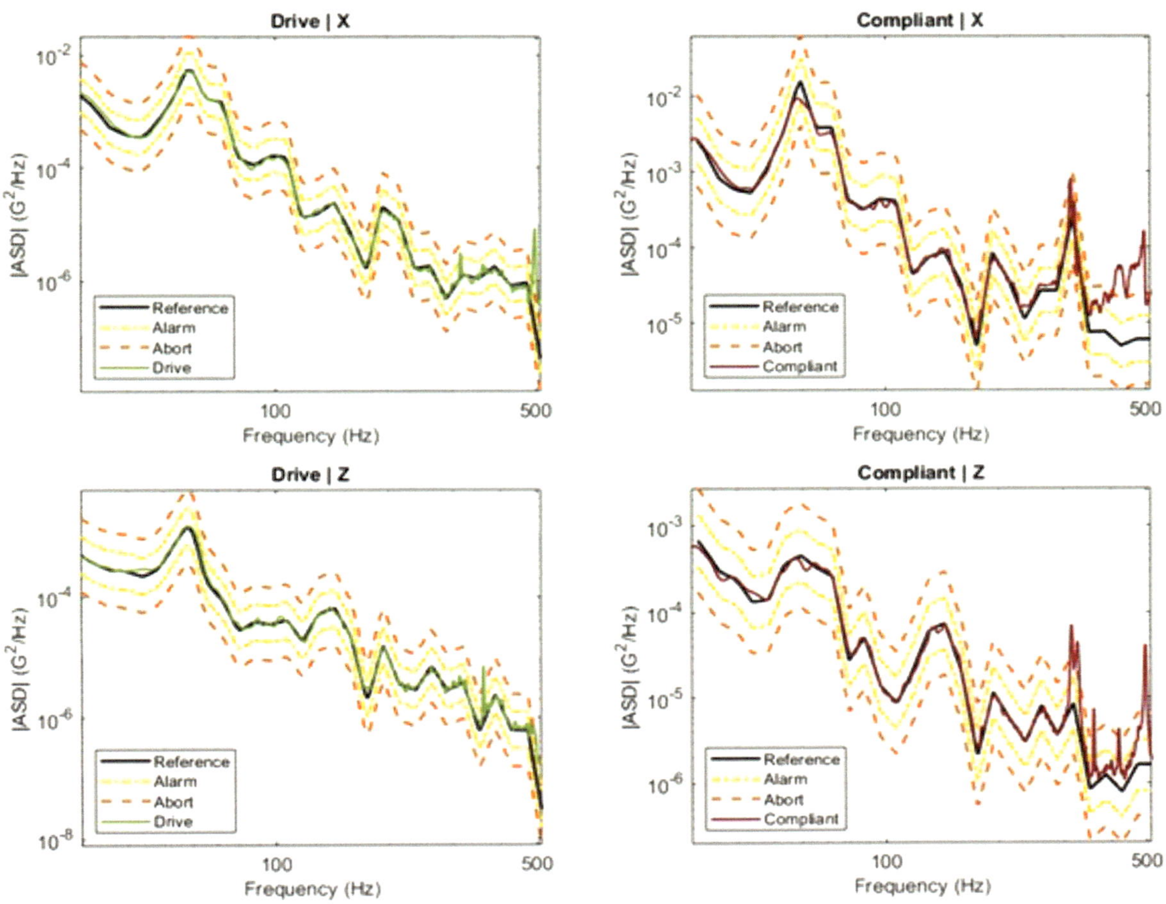

Fig. 4 Response at controlled DOFs compared to reference specifications derived from field data with ± 3 dB alarm limits and ± 6 dB abort limits.

the drive DOFs for the entire frequency range. When we look at numerical assessment of the control DOFs independently of the figure in 1, we can see that based on G_{RMS} alone, the compliant control appears the perform better, but when comparing the dB error and PSDPM, we see that the drive control error values were lower. This assessment is in alignment with the visual inspection of the control ASD which confirms that much of the response was outside the alarm and abort lines after 300 Hz in the compliant control which can be seen in 4. This excursion is not reflected in the G_{RMS} values but is reflected in the e_{dB} and PSDPM. The excursion above 300 Hz could be due to the low amplitude level of the specification, or simply that the controller was not able to match the specification in that region, especially in the Y direction. Already we can see that the controlling at the drive location was more achievable for the controller, however in further sections we will explore how the control affects the other responses in the system.

Table 1 Control Channel Error Metrics

Control Channel Error Metrics				
Configuration	**DOF**	**G_{RMS}**	**dB Error**	**PSDPM**
Reference Drive	X	0.291	0.0	1.0
	Z	0.155	0.0	1.0
Drive	X	0.283	0.508	0.925
	Z	0.157	0.678	0.921
Reference Compliant	X	0.439	0.0	1.0
	Z	0.134	0.0	1.0
Compliant	X	0.417	11.6	0.137
	Z	0.134	2.650	0.695

Next, we will examine the results using metrics of different fidelity levels to determine if we can tell which control DOFs resulted in a better matched global response to the field data and at what fidelity in response we are able to arrive at this conclusion.

Total ASD

At the highest level, we consider the sum of the 21 channels in the field test compared to the sum of the channels in the laboratory test. Before assessment, the G_{RMS} was used to perform a statistical t-test at 90% confidence to ensure there was a statistical difference between the field response and the environmental test responses. The t-test showed that there was a significant difference in the global response. In Figure 5, we see that for both the compliant and drive control configurations,

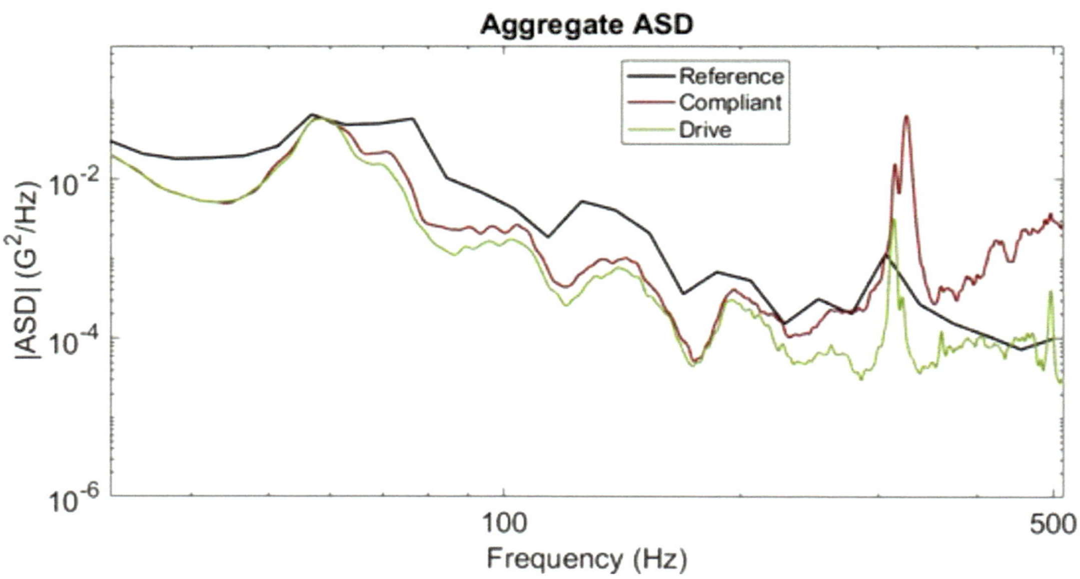

Fig. 5 The sum of all ASDs from the field test (reference), the drive control configuration, the compliant control configuration.

the response appears below the field response for most of the frequency range. However, after a large excursion at 350 Hz, the compliant control configuration has more energy than the field reference. This region is the same one where we noted that the controller was not able to match the specification, shown in Figure 4, which could be the reason for the excess energy in the aggregate ASD.

In Table 2, we can see that compliant configuration resulted in a G_{RMS} closer to that of the reference. We notice high dB error for the drive DOFs but a better PSDPM than the compliant control. Overall at this stage of assessment, it is difficult to tell which control DOFs performed closer to the field reference.

Table 2 Error Metrics for aggregate ASD response

Aggregate ASD Error Metrics			
Configuration	**G_{RMS}**	**dB Error**	**PSDPM**
Reference	1.562	0.0	1.0
Drive	0.987	-4.857	0.430
Compliant	1.390	1.930	0.387

X, Y, Z ASD

Next, we will assess how the control DOFs affected the total response in the principal axes of the system, X, Y, and Z. Note that there were several sensors that were not placed on the principal axes, and these channels were projected onto the principal axes using transformation matrices. In Figure 6, we see that the combined responses in the X-direction appear to follow the field reference well. However, the drive configuration is just below the specification for the entire frequency range of interest while the compliant direction follows the reference much more closely. Both configurations show a high response near 320Hz. In the Y-direction, the response do not match the specification well, most likely because there were no control DOFs in the Y-axis. In the Z-direction, both configurations appear to follow a somewhat similar profile of the specification but at a lower magnitude until near 320 Hz. This result is surprising because the Z-direction corresponds to up and down

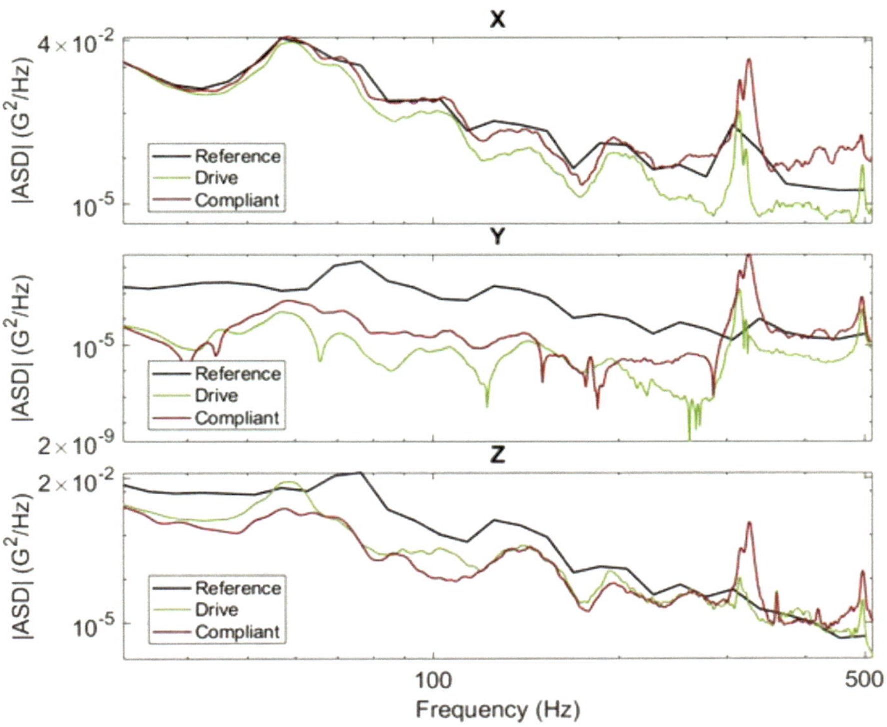

Fig. 6 The sum of the X, Y, and Z direction ASDs from the field test (Reference), the drive control configuration, and the compliant control configuration.

movement of the truck bed which was found to be mostly rigid through the assembly. As such, we would expect that having control channels in that direction would result in a combined response more similar to the field.

Numerically, we see that the drive location has a larger magnitude dB error in all directions compared to the compliant control. In addition, the PSDPM for the compliant control was higher in the X-direction, indicating that it was better agreement with the reference. However, the Z-direction PSDPM was slightly higher for the drive configuration than the compliant.

Table 3 Error metrics for combined X, Y, and Z ASD Response

Control Channel Error Metrics				
Configuration	**DOF**	**G_{RMS}**	**dB Error**	**PSDPM**
Reference	X	0.948	0.0	1.0
	Y	0.628	0.0	1.0
	Z	0.913	0.0	1.0
Drive	X	0.756	-5.657	0.304
	Y	0.112	-15.00	0.180
	Z	0.518	-2.50	0.582
Compliant	X	0.985	2.314	0.609
	Y	0.535	-6.791	0.394
	Z	0.394	-1.746	0.472

Individual channels

Finally, we will look at single response channel to assess the global response of the system. In an environmental test, looking at each monitored, response accelerometer could be quite time consuming and would be considered a last resort as an assessment of a test's quality. However, we must check that individual responses agree with conclusions that we may draw from the previous, coarse analyses, so we can access the reliability of the summed ASDs.

We will assess the individual accelerometer response at location 1504. Location 1504 is on the top surface of the cylinder of the GOBLET, at approximately 180 degrees from the drive location and 90 degrees from the compliant location as shown in Figure 7. This location was chosen because it is not on the same part of the assembly as the control DOFs for either configuration. Therefore, location 1504 is representative of a common practice in environmental testing where it is assumed

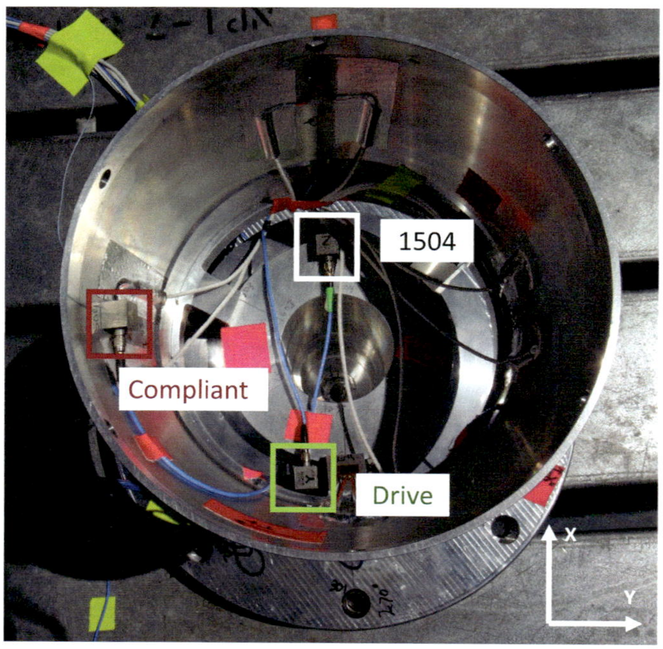

Fig. 7 Location 1504 with respect to the drive and compliant locations.

that controlling an assembly at one location results in a response similar to the field environment across portions of the assembly that are in proximity to the controlled DOFs.

The X DOF appears to follow the field response well, especially in the drive configuration. The compliant configuration follows the field response profile well until about 250 Hz which is near where the controller was not performing well. The Y DOF is much lower than the field response, which was expected as there was no control DOF in that direction. The Z direction, which had a control DOF, appears to not follow the field response as closely as the X direction, but has a similar profile and results in a slight over test of the the 1504 location. These results appear consistent with the total response and the principal axes response.

Assessing the error metrics in 4, we can similar difference in the G_{RMS} in both the configurations. We see low dB error for the Z-direction response for both configurations. The PSDPM for the drive configuration is much higher than the compliant configuration in the X-direction but is similar, if not lower, in the Y and Z-directions.

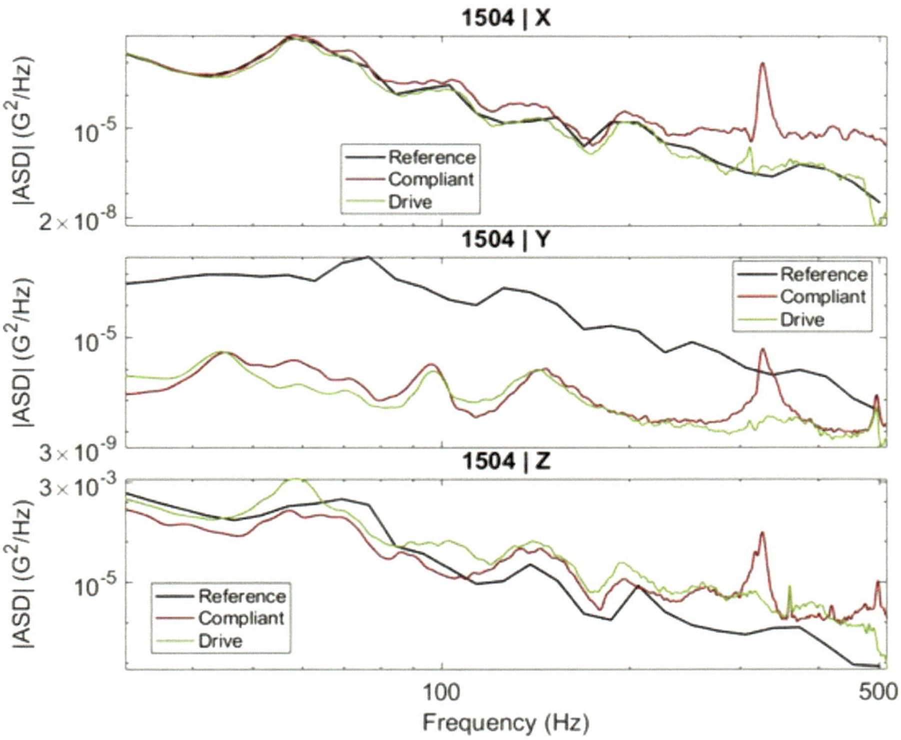

Fig. 8 Response of single accelerometer placed on cylinder portion of the assembly from the field (reference), the drive control configuration, and the compliant control configuration.

Table 4 Error Metrics for Response at Location 1504

Control Channel Error Metrics				
Configuration	**DOF**	**G_{RMS}**	**dB Error**	**PSDPM**
Reference	X	0.314	0.0	1.0
	Y	0.306	0.0	1.0
	Z	0.314	0.0	1.0
Drive	X	0.285	-0.744	0.739
	Y	0.008	-22.27	0.103
	Z	0.242	5.98	0.260
Compliant	X	0.352	8.23	0.439
	Y	0.012	-19.4	0.182
	Z	0.146	5.870	0.330

Conclusion

The goal of this study was to determine if there were statistically significant differences in the global response of the GOBLET when it was tested using two different sets of control DOFs. In addition, we examined how a test engineer might judge the quality of a test at different levels using a several error metrics. Overall, combining visual and numerical assessments, it is difficult to tell which control configuration performs better. Visual assessment might lead one to the conclusion that the drive configuration performed better, but numerical assessment at different levels showed mixed results. At the total aggregate level, the dB error and G_{RMS} would suggest the compliant control performed better than the drive, but the PSDPM would suggest otherwise and visual inspection suggests they were similar. At the next level, the sum of each direction, the compliant configuration also appeared to perform closer to the field in all error metrics. The individual channel assessment appears that the drive configuration performed better than the compliant due to its lower dB error, especially in the X-direction and higher PSDPM values.

Overall, we learn that at different levels, visual inspection and error metrics did not give a clear picture of which control configuration performs "better." This conclusion shows that aggregate assessment techniques do not necessarily provide similar answers when comparing data at different levels of fidelity, showing that there is need to make a full assessment of each response. The importance of metric choice also cannot be overstated, with several of the metrics in this paper often showing disagreement. The disagreement between metrics highlights that metrics are application specific and should be used with intention. While we were not able to conclusively determine which control configuration performed the best, this study highlights the need for exercising skepticism when analyzing aggregate ASD response and for assessing the error in response using multiple, relevant techniques to get a clearer picture of what the data are showing.

Acknowledgments This study was funded by the Los Alamos Delivery Environments Program. This work has been reviewed for public release (LA-UR-24-30418).

References

1. Daborn, P. *Smarter Dynamic Testing of Critical Structures*. PhD thesis, University of Bristol, 2014.
2. Rohe, D.P., Nelson, G.D., and Schultz, R.A. "Strategies for shaker placement for impedance-matched multi-axis testing". In Walber, C., Walter, P., and Seidlitz, S., editors, *Sensors and Instrumentation, Aircraft/Aerospace, Energy Harvesting & Dynamic Environments Testing, Volume 7*, pages 195–212, Cham, "February" 2020. Springer International Publishing.
3. Cross, K., Melendez, G., and Zwink, B. "Natural excitation within a test frame experimental results.". *Proceedings of the 41st International Modal Analysis Conference*, February 2023.
4. Smallwood, D.O. and Gregory, D.L. "Evaluation of a six-dof electrodynamic shaker system.". Technical report, Sandia National Lab.(SNL-NM), Albuquerque, NM (United States), 2008.
5. Kramer, H., Schultze, J., Danforth, S., and Mann, B. "Quantifying differences between mimo and siso testing on the marc structure". In *Proceedings of the 42nd International Modal Analysis Conference*, February 2024.
6. Danforth, S., Johnson, S., Fenstermacher, K., Ramirez, A., Sedillo, H., Schultze, J., and Ferguson, E. "Developing methods for single-axis base excitation vibration testing with additional distributed shakers". In *31st International Conference on Noise and Vibration*, September 2024.
7. Lee, D., Ahn, T.S., and Kim, H.S. "A metric on the similarity between two frequency response functions". *Journal of Sound and Vibration*, 436:32–45, September 2018.

Chapter 12
Estimating Component Level Environments Using Next Assembly Measurements

Marcus Behling, Matthew S. Allen, Randall L. Mayes, and Washington J. DeLima

Abstract A significant obstacle to properly recreating vibration-induced damage in qualification tests is the lack of availability of flight environment data. Typically, data from a previous flight with similar hardware is used to create the specification for the new flight. In addition, data is usually only available on the next assembly and not on the components that are being qualified, though in-flight component damage is more accurately reproduced when the environment is available on the component itself. Hence, this study seeks to estimate the flight environment on the device under test (DUT) using environment data only on the next assembly as only this data is typically available. This is performed using the System Equivalent Reduction Expansion Process (SEREP). The accuracy of the response expansion is shown to be sensitive to the amount of next assembly used to model the flight system's dynamics as well as the condition number of the mode shape matrix. Full flight system dynamics are shown to not generally be observable with accelerometers on the next assembly only, so some error is expected in DUT response estimation with only next assembly measurements. The condition number of the mode shape matrix is shown to improve significantly when a few measurements are available on the DUT, so accurate full-field DUT responses can be estimated if the next assembly is well instrumented and if there are a few measurement channels on the DUT. Hence, one can significantly improve damage simulation if sparse DUT environment measurements are available.

Introduction

Traditional single axis vibration tests generally result in improper simulation of operational vibration induced damage [1][2][3][4]. These tests are usually assumed to cause more damage than would occur in the actual flight due to enveloping of the environment and an impedance mismatch between the test fixture and in-flight next assembly [5]. Off-axis motion, e.g., due to rotational shaker dynamics, can also artificially increase single-axis test severity [6]. Impedance Matched Multi-Axis Tests (IMMATs) were developed to fix some of the problems with traditional tests [7]. They use multiple shakers to simultaneously excite multi-axial motion and part of the next assembly is included in the lab test to match the flight boundary condition better. These tests have shown the potential for improved environment reconstruction, particularly when there are more control accelerometer channels than shakers and when these accelerometers are well distributed across the DUT [8].

However, environment data is typically not available on the DUT, and if it is, there are usually not enough accelerometers to completely capture the dynamics of the component. More often, rather, environment data is only available on the next assembly from a past flight. The usual practice is to mount the component on a large, single-axis shaker system and apply the

Marcus Behling
Graduate Research Assistant, Brigham Young University – Mechanical Engineering Department
e-mail: mebehlin@student.byu.edu

Matthew S. Allen
Professor, Brigham Young University – Mechanical Engineering Department
e-mail: matt.allen@byu.edu

Randall L. Mayes
Mayes Consulting
e-mail: rajnd75@msn.com

Washington J. DeLima
Honeywell Federal Manufacturing & Technologies
e-mail: wdelima@kcnsc.doe.gov

Alexandra Karlicek et al. (eds.), *Dynamic Environments Testing, Vol. 7*, Conference Proceedings of the Society for Experimental Mechanics Series, https://doi.org/10.13052/9788743801528-12

next assembly motion one axis at a time to the base of the component. The authors proposed a method that seeks to improve on this, which is called the Transmission Simulator IMMAT Framework (TS-IMMAT). This method uses multiple shakers to cause the base of the component to exhibit the same motion that it did during flight. If the base of the component is rigid then this entails using at least six shakers to control to six degrees of freedom, although additional shakers can be used if the base is flexible. While this can improve environment reconstruction relative to a traditional single-axis test, past experience has shown that these tests do not generally reproduce the environment on the component as accurately as an IMMAT test [9]. Hence, if one could use the response on the next assembly to estimate the response on the DUT, then one could perform an IMMAT test rather than a TS-IMMAT test, and reproduce the response on the DUT more faithfully, reducing the costs associated with failed parts due to over-test or in field failures due to under-test. Additionally, some tests require component level data, e.g., the intrinsic connection excitation (ICE) framework [10] and fixture neutralization (FINE) methods [11]. In general, controlling component level data compensates for boundary condition mismatches between the flight and lab setups and is useful [12].

The goal of this study is to see if accurate DUT response reconstruction is possible given next assembly measurements only. Specifically, we hope to answer: 1) How much of the next assembly should be included in the expansion model to accurately reconstruct the response on the DUT? 2) How accurate can one expect the reconstructed DUT responses to be if only next assembly measurements are available? 3) If next assembly measurements are insufficient, what is the minimum number of accelerometers that can be placed on the DUT to improve response estimation at unknown locations? These questions are studied by analyzing the Finite Element Model (FEM) case study shown in Figure 1, where the DUT is a stool intended to represent a structure on which a circuit board might rest, and the next assembly is a wedding cake-like structure. In this study, it is assumed that the next assembly geometry does not change from when the environment is measured (or simulated, in this case), i.e., the same system that data was measured on is the current system of interest. In practice, data may only be available from a previous flight, so the next assembly geometry may change. In this case, the methods presented in [13] might be applied to estimate the environment on the new next assembly, and the methods presented in this paper could then be applied to estimate the DUT environment.

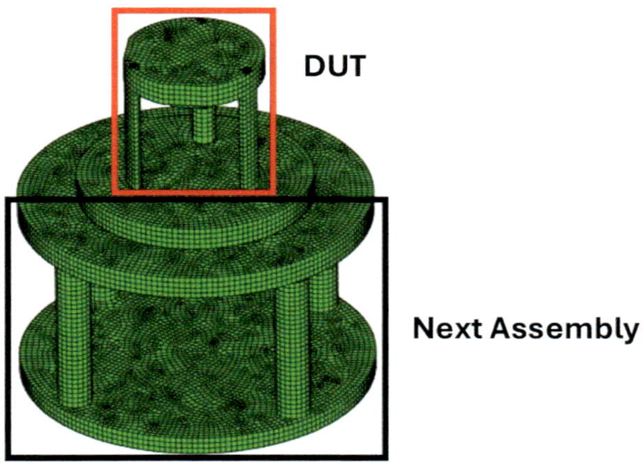

Fig. 1 Flight Setup Case Study

Environment Estimation Methods

The first efforts to accurately estimate flight environments in the absence of measured data seem to have been at NASA in the 60's. One study presented two methods, one for components directly excited by acoustic or aerodynamic pressures, and another for those primarily excited by transmission through other parts of the structure. These methods involved scaling previously measured responses using ratios of forcing function magnitude and the relative masses of the original and new component [14]. Another study divided environment estimation methods into two categories: gross prediction techniques, in which broad correlations between forcing function and response magnitude are derived from previous flights, and custom prediction techniques, which involve more detailed modeling of the forcing function and structure of interest [15]. Statistical energy methods were also proposed as a way to predict environments at higher frequencies [16]. In short, early methods generally modified previous flight data to estimate new environments, though they were hindered by a lack of computational power and the inability to create accurate models of system hardware. These methods might accurately estimate response magnitude, though it would be difficult to apply them and obtain a physically meaningful environment at multiple locations on a DUT.

More recent methods benefit from improvements in computational power and the ability to model systems accurately via modal testing and FEM. Various response expansion methods have been devised, and some of these are summarized in [17]. Perhaps the most well-known is the System Equivalent Reduction Expansion Process (SEREP), in which the system modal responses are obtained from a sparse set of measurements and then expanded back out to estimate responses at desired degrees of freedom [18]. Another is the pseudo-force method, in which a set of forces is applied in simulation to the system of interest, the FRF relating these forces to the known responses is inverted, and the forces are applied to the system to obtain responses at the desired degrees of freedom [19]. Chebyshev polynomials have also been used as the basis functions that responses are projected onto, though this may not extend well to complex geometries [20]. Each of these methods can be used to obtain accurate responses at various locations on a DUT if the basis functions being projected onto – e.g., mode shapes, frequency response functions, or Chebyshev polynomials – span the DUT's response space and are well conditioned.

Conditions for Accurate Response Estimation

The response expansion method of choice in this study is SEREP. SEREP was introduced in [18], although the idea of using the modes of a structure as a basis to estimate or control the motion or stresses in a component is well known and used in many areas without reference to the formalisms put forth in SEREP (see, e.g. [21]). The SEREP theory is briefly rewritten here in the context of dynamic environment testing specifically. In flight, the system response is a linear combination of some set of flight mode shapes times their response amplitudes as shown in Equation (1),

$$\mathbf{x}_m^{\mathrm{DUT+NA+V}}(\omega) = \mathbf{\Phi}_m^{\mathrm{DUT+NA+V}}\mathbf{q}^{\mathrm{DUT+NA+V}}(\omega) \tag{1}$$

where the superscript indicates that the DUT and Next Assembly (NA) are installed in the flight Vehicle (V), and the subscript indicates that the response is observed at a set of points m. Note that the set of points m may include points on the next assembly, denoted p as well as points on the DUT, denoted d, so $m = p \cup d$ (or, in Matlab notation, $m = [p; d]$). A full FEM of the flight vehicle is often not available, and if it is, it is typically not detailed enough to accurately represent the dynamics of the DUT and its next assembly locally. Instead, we assume that a FEM is available of the DUT attached to the Next Assembly but with free boundary conditions. We also assume that the mode shapes of this assembly accurately span the response space of the DUT and Next Assembly in flight as in Equation (2),

$$\mathbf{x}_m^{\mathrm{DUT+NA+V}}(\omega) \cong \mathbf{\Phi}_m^{\mathrm{DUT+NA}}\mathbf{q}^{\mathrm{DUT+NA}}(\omega). \tag{2}$$

There is certainly no guarantee that the mode shapes of the DUT and Next Assembly in the free configuration will span the space of the motions observed when they are assembled into the vehicle. There may be dynamic interaction between the Next Assembly and the rest of the vehicle, and the connection to the vehicle may load the interface, altering the response. Past studies have sought to understand how much of the flight system should be included in the Next Assembly [22][23]. There is no general answer, so best judgment must be employed to include the portions of the vehicle adjacent to the DUT that significantly influence its dynamics and to exclude insignificant portions.

Assuming a set of accelerometers was installed on the next assembly in flight, SEREP can be applied to a set of measured responses to estimate the modal responses of the DUT + Next Assembly configuration using the Moore-Penrose pseudoinverse (denoted $()^+$ below). For this calculation, we assume that a set of measurement points p are available and that these measurements are only available on the Next Assembly. Hence p would be a subset of the available measurement points m considered in Equation (2),

$$\mathbf{q}^{\mathrm{DUT+NA}}(\omega) = \left(\mathbf{\Phi}_{\mathrm{p}}^{\mathrm{DUT+NA}}\right)^+ \mathbf{x}_{\mathrm{p}}^{\mathrm{DUT+NA+V}}(\omega). \tag{3}$$

If the modal responses are accurately estimated, then the physical responses of the DUT, denoted points d can be estimated using the mode shapes at the desired DUT degrees of freedom as in Equation (4),

$$\mathbf{x}_{\mathrm{d}}^{\mathrm{DUT+NA+V}}(\omega) \cong \mathbf{\Phi}_{\mathrm{d}}^{\mathrm{DUT+NA}}\mathbf{q}^{\mathrm{DUT+NA}}(\omega). \tag{4}$$

As previously mentioned, accurate response reconstruction partially depends on satisfying Eq. (2), i.e., selecting the next assembly so that the modes of the DUT + Next Assembly in the free configuration (i.e. when not installed on the vehicle) span the space of their response in flight. In addition, the DUT + Next Assembly modes must be observable with the given accelerometer set p, and the mode shape matrix, $\mathbf{\Phi}_{\mathrm{p}}^{\mathrm{DUT+NA}}$, must be well-conditioned. When the condition number of this matrix is high, any noise in the flight environment will cause significant error in estimating the modal responses, $\mathbf{q}^{\mathrm{DUT+NA}}(\omega)$, resulting in error in estimating the DUT's flight environment, $\mathbf{x}_{\mathrm{d}}^{\mathrm{DUT+NA+V}}(\omega)$. To accurately estimate the response using this approach one must have at least as many accelerometers as there are active system modes and they must be placed to effectively distinguish those modes.

In general, though, there may be modes of the combined DUT and Next Assembly system that are not observable with accelerometers on the next assembly only. When this is the case, some environment data is required on the DUT, so some of the sensors d must be added to Equation (3). To determine how many accelerometer channels are required on the DUT to fully observe all system modes, the Singular Value Decomposition (SVD) can be applied to the mode shape matrix, $\mathbf{\Phi}_{\mathrm{p}\cup\mathrm{d}}^{\mathrm{DUT+NA}}$, as in Equation (5),

$$\mathbf{\Phi}_{\mathrm{p}\cup\mathrm{d}}^{\mathrm{DUT+NA}} = \mathbf{U\Sigma V^T}, \tag{5}$$

where $\mathbf{\Sigma}$ is an $\mathrm{n_{accels}}$ x $\mathrm{n_{modes}}$ diagonal matrix with singular values on the diagonal in order of descending magnitude. A normalized singular value vector, $\mathbf{\Sigma}_{\mathrm{norm}}$, can be calculated by dividing each singular value by the largest singular value as in Equation (6),

$$\mathbf{\Sigma}_{\mathrm{norm}} = \left\{ \ \sigma_1/\sigma_1 \quad \sigma_2/\sigma_1 \quad \ldots \quad \sigma_{\mathrm{n_{modes}}}/\sigma_1 \ \right\}^{\mathrm{T}}. \tag{6}$$

If there is some maximum acceptable condition number, $\mathrm{C_n}$, of the mode shape matrix $\mathbf{\Phi}_{\mathrm{p}\cup\mathrm{d}}^{\mathrm{DUT+NA}}$, which is inverted in the projection process, then the number of entries in $\mathbf{\Sigma}_{\mathrm{norm}}$ greater than the reciprocal of $\mathrm{C_n}$ is the number of modes that are observable with the next assembly accelerometer set, $\mathrm{n_{mo}}$, and the number of entries less than the reciprocal of $\mathrm{C_n}$ is the number of modes that are not observable with the next assembly accelerometer set, $\mathrm{n_{mu}}$. At least $\mathrm{n_{mu}}$ accelerometer channels must be measured in flight on the DUT, therefore, to observe all modes of the DUT. In practice, it is likely that a few more channels are required than this minimum to reduce the condition number of the mode shape matrix to an acceptable value.

In the following sections, this theory is applied to the system in Fig. 1 for three separate cases. In the first case, a small portion of the next assembly is included in the projection model, and the environment is assumed to only be known on the next assembly. In the second case, the only available measurements are still on the next assembly, though the full in-flight next assembly is included in the projection model. Finally, in the third case a few accelerometer channels are included on the DUT with the full next assembly also included. These case studies are summarized in Table 1.

One final note is that tests are practically performed using Power Spectral Densities (PSDs) rather than linear spectra; these two can be related according to,

$$\mathbf{S}_{\mathrm{XX}}(\omega) = \mathrm{E}\left(\mathbf{x}(\omega)\,\mathbf{x}(\omega)^*\right), \tag{7}$$

where "E" is the expected value operator, or the average over many measurements.

Table 1 Definition of Cases Studied

Case	Assembly Considered	Measurements Available
Case 1	DUT+small portion of the next assembly	on next assembly only
Case 2	DUT+full in-flight next assembly	on next assembly only
Case 3	DUT+full in-flight next assembly	on next assembly and DUT

Results

A flight environment was simulated on the flight vehicle setup in Fig. 1 by applying a 1 $\mathrm{N^2/Hz}$ force from 10-2000 Hz at each of the locations shown in Figure 2. An FRF relating the forces at these locations to the responses at the desired locations was synthesized using all 13 of the system's modes below 2000 Hz. A damping ratio of 0.01 was assumed for each of the elastic modes. This environment is used for each of the three cases presented here.

Case 1: Small portion of next assembly, no DUT data

In the first case, only the top plate of the next assembly in Fig. 1 is included in the model. Hence, while the flight assembly includes all of the components shown in Figure 1, the lab assembly includes only the components shown in Figure 3. This represents a case where the full flight vehicle model is not available, so a portion of the supporting structure is included to represent the local dynamics of the vehicle. There are 9 modes of this system below 2000 Hz that were included in the

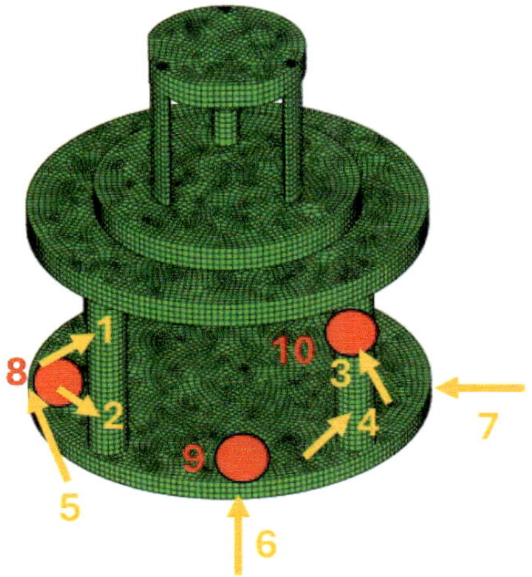

1,3 Applied to Columns in Radial Direction
2,4 Applied to Columns in Spin Direction
5 Applied to Side of Bottom Plate in Spin Direction
6,7 Applied to Side of Bottom Plate in Radial Direction
8,9,10 Applied to Bottom of Bottom Plate in Launch Direction
4 Additional Forces (2 radial, 2 spin) applied to back columns (Similar to Locations 1 through 4)

Fig. 2 Force Locations Used to Simulate Flight Environment

model, and a damping ratio of 0.01 was again assumed for these modes. The accelerometer locations that are projected onto (labeled 1-5), as well as the reference accelerometer 6 on the DUT that we hope to recreate the response of, are shown in Figure 3. All accelerometers are triaxial, measuring the response in a global xyz coordinate system where the z axis is parallel to the columns and will be referred to as the launch direction while x and y are in the plane of the plate and will be referred to as the radial and spin directions. Hence, the mode shape matrix in Eq. (2) was 15×9. A small amount of noise is added to the flight environment to make this analysis more realistic and to show the effects of a high mode shape matrix condition number.

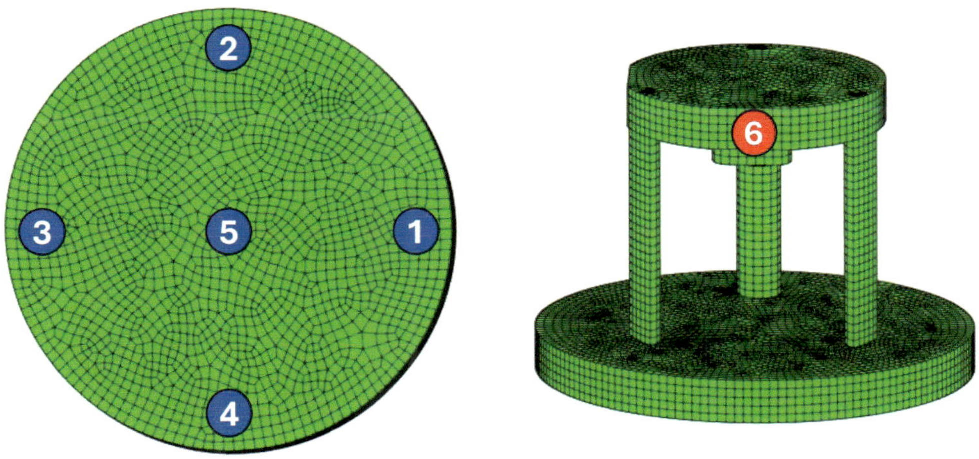

Fig. 3 Case 1 Assembly and Triaxial Accelerometer Locations (all on the plate) used in SEREP (Left) and Reference Triaxial Accelerometer on DUT (Right)

The stool environment estimated obtained from SEREP is compared to the actual flight environment in Figure 4, and there is clearly significant error in estimating the DUT responses. In this case, the estimated environment exceeds the magnitude of the actual environment at most frequency lines, so a test performed controlling to the estimated environment would result in high levels of over-testing. Conservatism is necessary in design and testing, though this level of conservatism seems extreme, and the actual test level would likely be increased due to enveloping, adjustments for increased reliability, and so on [24]. So, even though the environment is generally conservative and even has similar natural frequencies to the actual environment, this estimate is considered insufficient.

One reason this estimate is so poor is that the mode shapes of the Stool-Plate configuration do not quite span the in-flight response space of the DUT. As previously mentioned, this configuration has 9 modes below 2000 Hz, while the flight configuration has 13. One of these flight modes, shown in Figure 5, is a bubble mode of the plate that is accentuated by

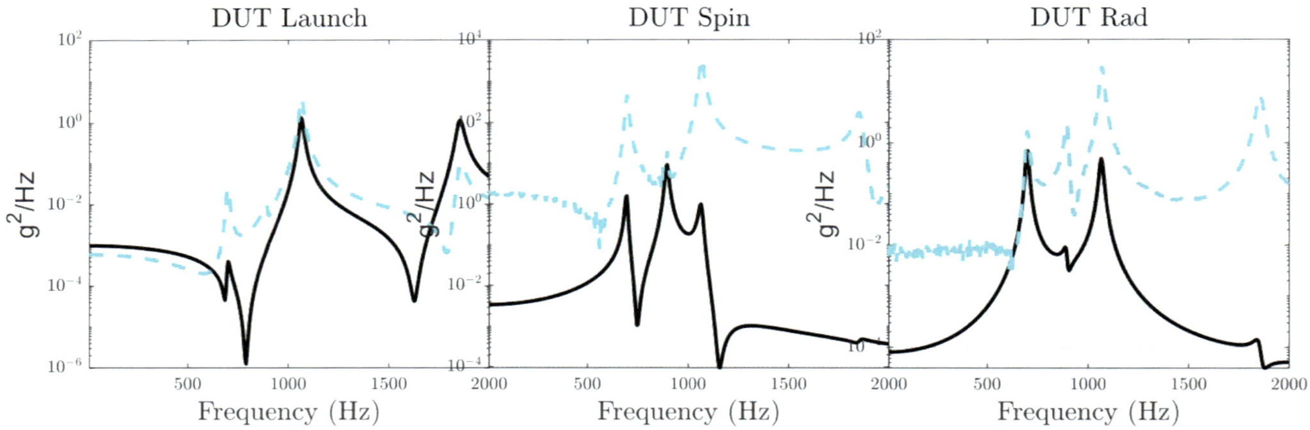

Fig. 4 Estimated DUT Response PSDs estimated by SEREP from noisy measurements for Case 1 (Cyan) vs. Actual Stool Response PSDs (Black)

Fig. 5 Flight Configuration Mode #13 at 1854 Hz

the columns, so it is not observed in the stool-plate configuration. This modal response is not properly replicated in the projection process, causing increased error in the DUT environment estimation.

Another reason for the inaccurate response estimation is that the modes of the Case 1 Stool-Plate configuration are not fully observable with the set of accelerometers used in the expansion process. Two of the three elastic modes of this configuration are shown in Figure 6; the third is a bending mode similar to the right side of Fig. 6 but in the out-of-page direction. In these modes, the stool experiences significant elastic deformation, while the plate remains mostly rigid. To the plate accelerometers, these modes are basically indistinguishable from the rigid body modes of the system. As a result, the responses of these modes cannot be accurately estimated from the measurements on the base plate, resulting in significant error in estimating their responses and hence the environment on the DUT.

To further demonstrate the inability of the plate accelerometers to fully observe the stool-plate system modes, the modal assurance criterion (MAC) was applied to the stool-plate mode shapes using the Case 1 accelerometer locations. A MAC value of 1 indicates that mode shapes are identical, while a MAC of 0 indicates that modes are orthogonal. Ideally, cross-MAC values between different modes should be very low, e.g., below 0.1, as this indicates that the set of sensors being used can properly distinguish between each mode and properly identify modal responses from environment data. As seen in Figure 7, though, the off-diagonal MAC terms are very high for the elastic modes. Specifically, the torsion mode is very similar to rigid body rotation about the z-axis, the first bending mode is similar to y-translation and x-rotation, and the

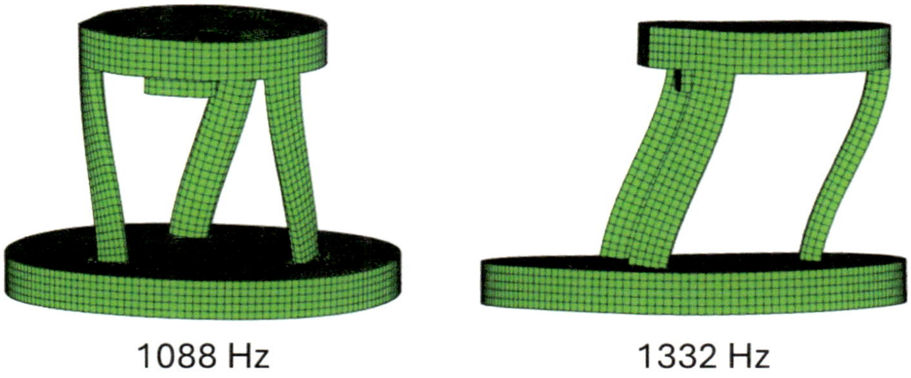

Fig. 6 Torsion and Bending Modes of the Case 1 Stool-Plate System

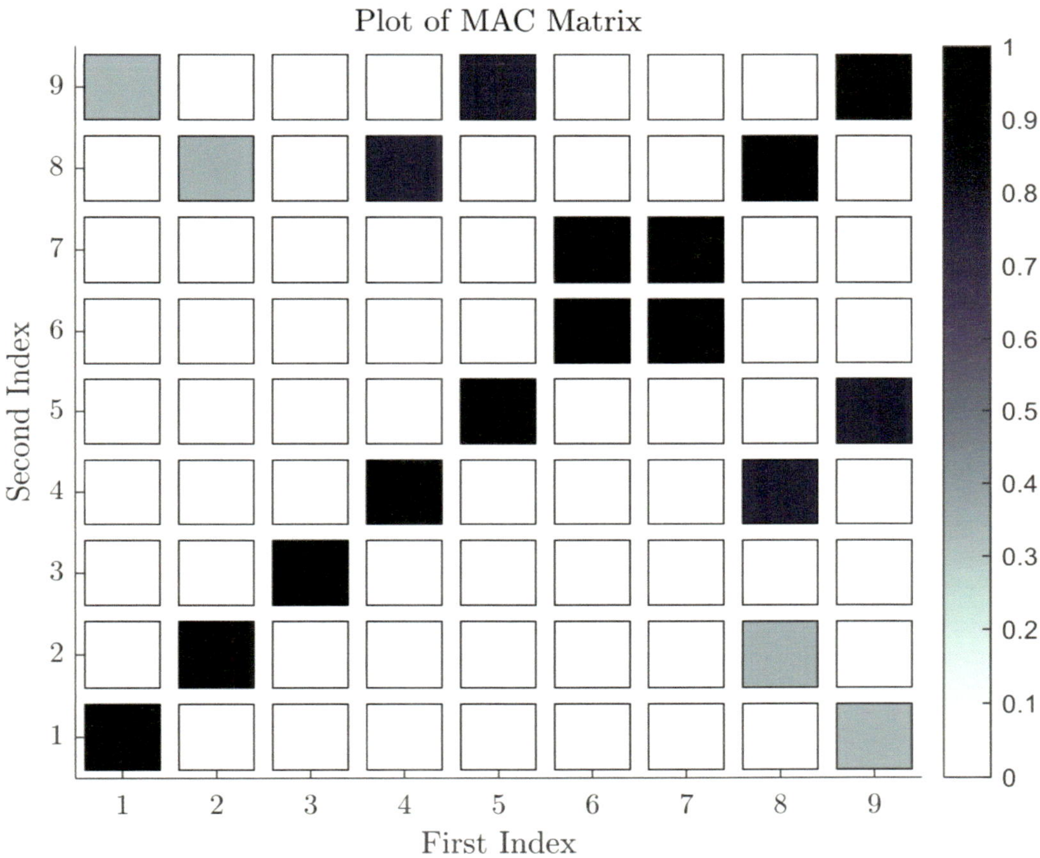

Fig. 7 MAC for Stool-Plate Configuration

second bending mode is similar to x-translation and y-rotation. This confirms that the elastic modes of the stool cannot be distinguished from the system's rigid body modes given the set of plate accelerometers that was used in Case 1 and that this is a cause of the error observed here.

Case 2 – Full next assembly, no DUT data

Since the portion of the next assembly included in the previous section did not accurately capture the motion of the flight assembly, the full-flight next assembly is used in the SEREP process in this case, although measurements are still assumed to not have been obtained on the DUT. The simulated accelerometer locations are shown in Figure 8. Since the number of channels (54) is much greater than the number of system modes (13), and since the accelerometers are well distributed across on the next assembly, we assume that the next assembly's motion is properly observed. The DUT reference accelerometer is the same as in the previous case.

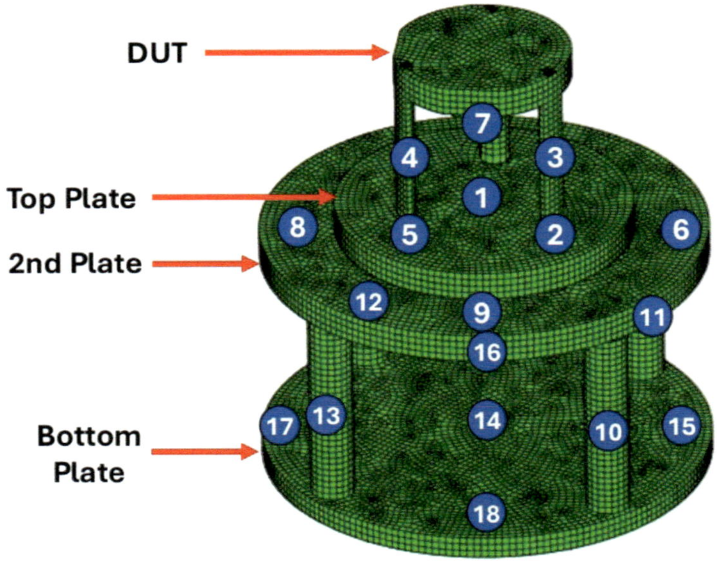

Fig. 8 Triaxial Accelerometer Locations Used in SEREP on Flight Configuration. 1-5 are on Top Plate, 6-9 are on 2nd Plate, 10-13 are on the Columns, and 14-18 are on the Bottom Plate.

As seen in Figure 9, the response reconstruction on the DUT is slightly improved over the first case due to the increased portion of the next assembly included, although significant error remains in the spin and radial directions. This suggests that the system's modes are not fully observable with Case 2 accelerometer set, which are on the next assembly only, and that the mode shape matrix must be ill-conditioned in this case.

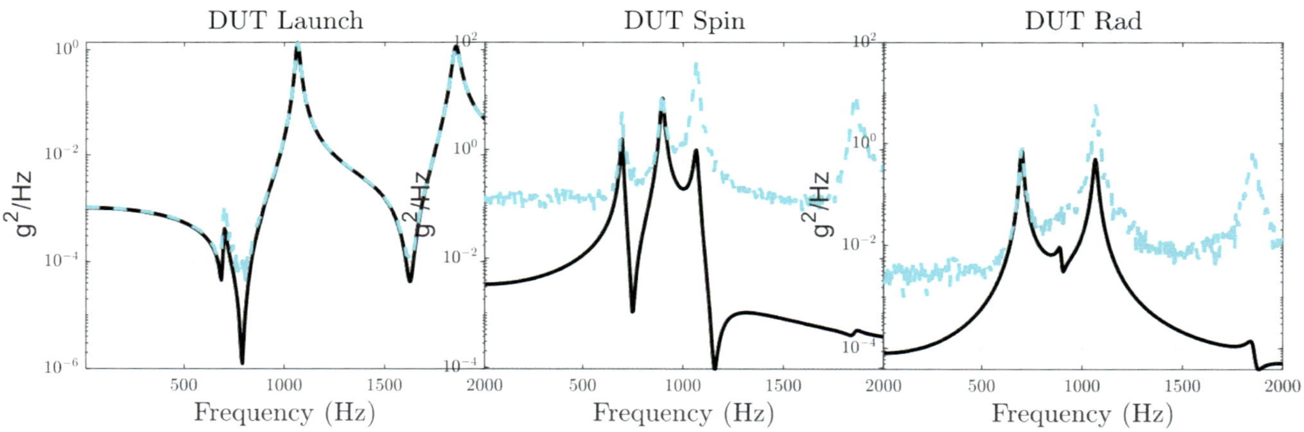

Fig. 9 Case 2 DUT Response Reconstruction Using SEREP with Noise in Measurements

The mode shapes of the flight system, shown in Figure 10, help explain these results. The 13th mode shape is not included in this figure, as it was shown previously in Fig. 5, and the 8th and 12th mode shapes are not shown because they are similar to the 7th and 10th modes in the orthogonal direction. More of the system's modes are identifiable with next assembly accelerometers in this case because the next assembly deforms along with the DUT. Modes 7, 8, 9, 11, and 13 seem to have unique shapes, although the next assembly's motion in modes 10 and 12 seems to be a combination of torsion and bending, so these modes are not observable with accelerometers on the next assembly only, and their modal responses are not expected to be accurately recreated.

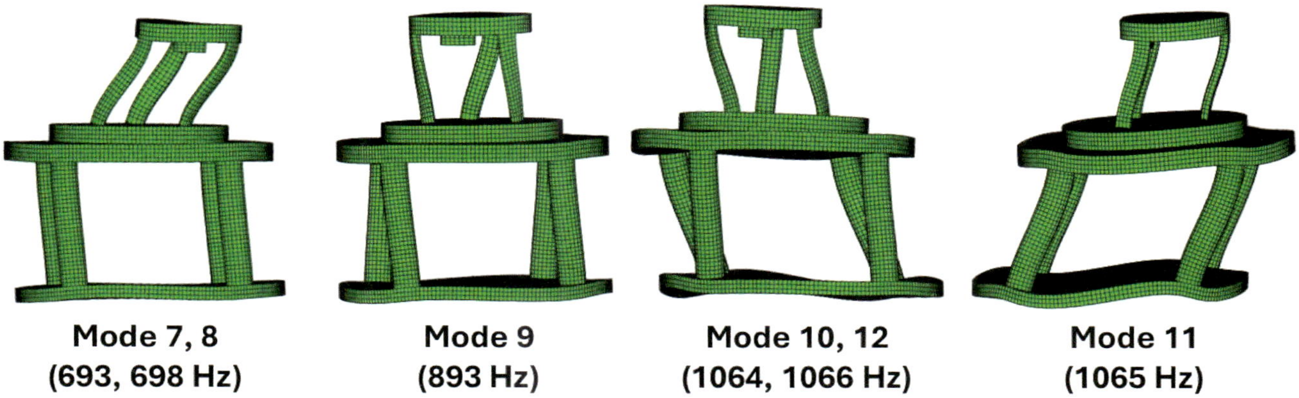

Fig. 10 Flight System Elastic Mode Shapes. Mode #13 is shown in Fig. 5.

These observations are confirmed by taking the SVD of the mode shape matrix using the accelerometer set employed in this case. The matrix $\left[\boldsymbol{\Phi}_{\mathrm{p}}^{\mathrm{DUT+NA+V}}\right]$ was 54×13, and the singular value ratios are shown in Table 2. Any singular value ratio less than 1% corresponds to a condition number of over 100. These seem to be the cause of the response reconstruction error in the studies performed here, and hence this seems to be a reasonable rule of thumb. With accelerometers on the next assembly only, 10 modes are observable, and 3 are not, so at least three accelerometer channels would be required on the DUT to properly observe these modes. This is assuming that there are no other candidate locations on the next assembly that would observe these modes, which seems to be a reasonable assumption given the number and spread of accelerometers that are already included on the next assembly.

Table 2 Normalized singular values of mode shape matrix for Case 2 and 3 assembly, comprised of the DUT + Full Next Assembly. These were computed without (Case 2) and with (Case 3) three DUT accelerometer channels.

Singular Value No.	No DUT Channels	3 DUT Channels
1	1	1
2	0.99	0.98
3	0.98	0.94
4	0.67	0.64
5	0.59	0.57
6	0.58	0.57
7	0.57	0.55
8	0.41	0.41
9	0.41	0.40
10	0.17	0.32
11	**6e-3**	0.07
12	**6e-3**	0.05
13	**2e-3**	0.03

Case 3 – Full next assembly with DUT data

Since next assembly accelerometers alone were unable to observe three of the flight system's modes, the environment would need to be known on at least three accelerometer channels on the DUT to properly estimate the environment at the unknown locations. The degrees of freedom shown in Figure 11 were added to the measurement set. The resulting 57×13 matrix $\left[\boldsymbol{\Phi}_{\mathrm{pUd}}^{\mathrm{DUT+NA+V}}\right]$ was found to reduce the condition number of the mode shape matrix significantly as shown in Table 2. Hence, with these three channels on the DUT and the previously shown channels on the next assembly, all system modes are observable, and all modal responses should be accurately estimated using SEREP, so the unknown DUT response in Fig. 3 should be estimated with little error.

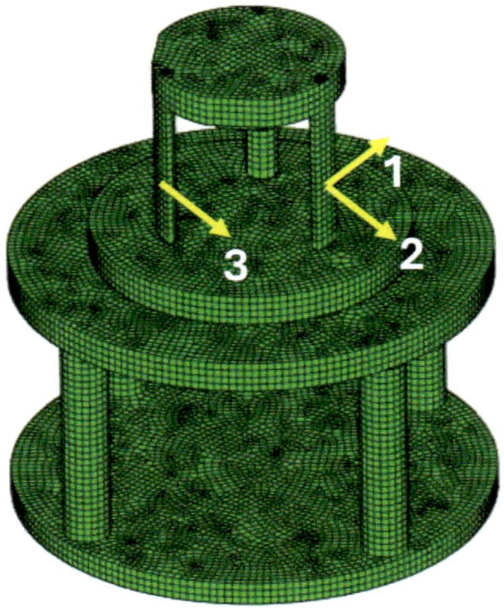

Fig. 11 DUT Accelerometer DOFs to Include in Modal Projection

As shown in Figure 12, this new accelerometer set yielded excellent reconstruction of the unknown DUT accelerometer environment. There is slight error in each direction, likely because the condition number of the mode shape matrix is $\sim$35, although it is small relative to the overall environment. The estimated environment is, again, conservative, so any error in estimation would not result in undertesting. Though only one accelerometer's response is shown here, we assume that the responses at any location on the DUT would be recreated with similar accuracy since this triaxial accelerometer's response is influenced by all of the modal responses. Hence, accurate replication of its response would require accurate replication of all modal responses, meaning that the DUT's response should be estimated accurately globally.

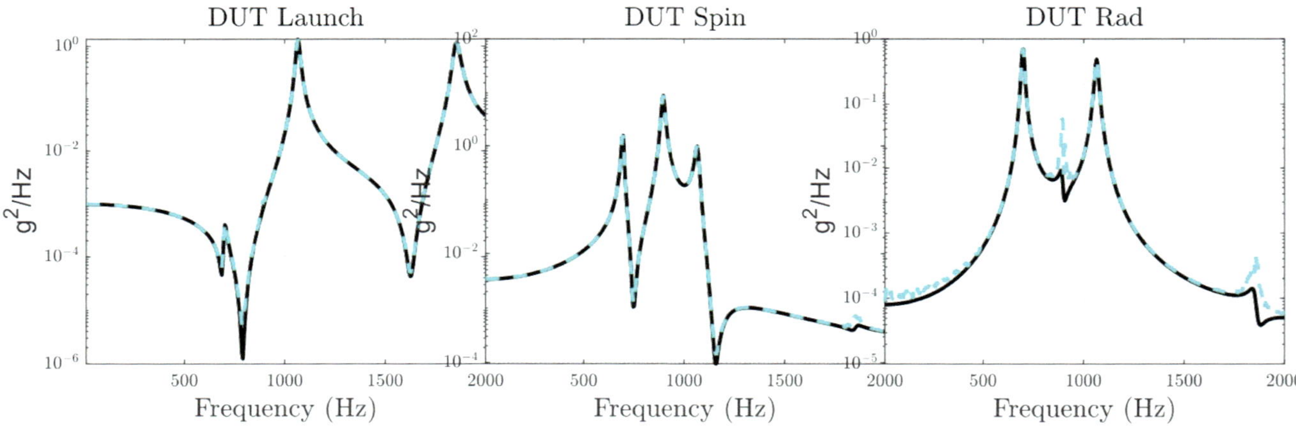

Fig. 12 Case 3 Reconstructed DUT Responses. Case 3 used the full Next Assembly plus three DUT channels.

The Benefit of Data on the DUT

In single axis tests, having component level data available on the DUT allows response limits to be implemented [5], which reduces the amount of overtest. In multi-input/multi-output (MIMO) tests, the DUT accelerometers can be controlled directly, resulting in improved reconstruction of the environment at uncontrolled locations on the DUT relative to controlling to measurements on the next assembly only. To show the benefit of having component level data, a MIMO test is simulated, first controlling the fixture (i.e. next assembly only) and then controlling the DUT responses. The shaker locations in Figure 13 are used for simulating the test. Though there are 13 modes of the system, 10 shakers at the locations in Fig. 13 were found to be sufficient for accurately reproducing the response.

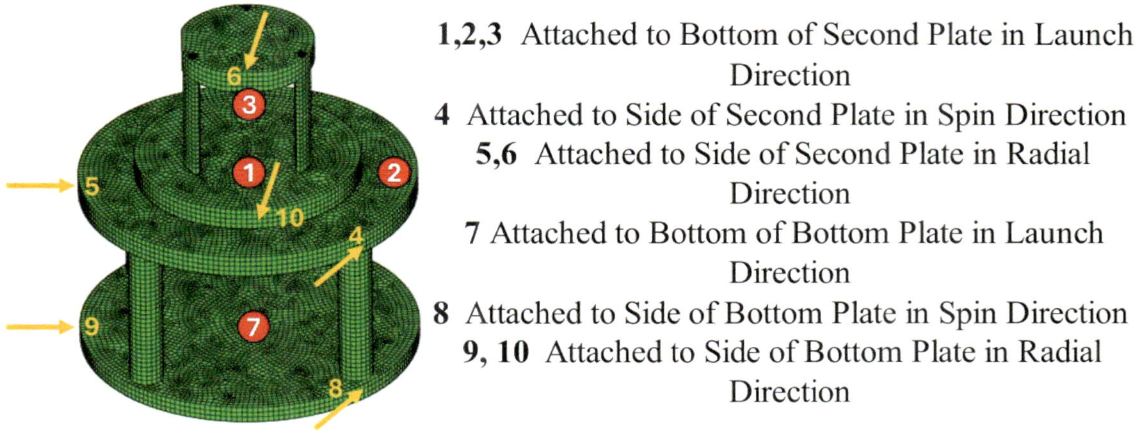

Fig. 13 Shaker Locations for Simulating MIMO Test

The next assembly control accelerometers are the same 54 channels used in Cases 2 and 3 and were shown in Fig. 8. The DUT control accelerometers are shown in Figure 14. In both the next assembly and DUT control cases, the goal is to replicate the environment at the reference DUT accelerometer shown in Fig. 3.

The environments reconstructed in the fixture and DUT control cases are compared to the original flight environment in Figure 15. When controlling only to accelerometers on the next assembly, there are noticeable errors in reconstructing the DUT's responses in the spin and radial directions. When the DUT accelerometers are controlled, however, there is almost no error in reconstructing the DUT environment. The control is improved because the DUT's motion is fully observed with the DUT accelerometer set, while the DUT's motion is not completely observed with accelerometers on the next assembly only. Hence, DUT environment data is useful because controlling it in MIMO tests results in improved simulation of vibration-induced damage relative to only having data on the next assembly and controlling the motion of the fixture. However, the reconstruction of the DUT environment is reasonable even with measurements on the next assembly only, suggesting that the shaker control may be less sensitive to noise and accelerometer locations than SEREP is. Hence, this may be a benefit to

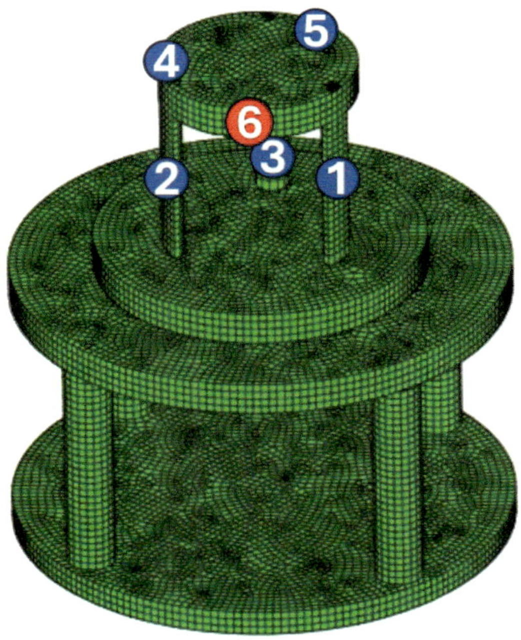

Fig. 14 DUT Control Accelerometer Locations. Accelerometers 1-3 are on the legs of the stool, while 4-6 are on the top plate. Reference Accelerometer in Red.

using the pseudo-forces method [19] instead of SEREP for environment estimation purposes, as the pseudo-force method is essentially the same as the control method used here.

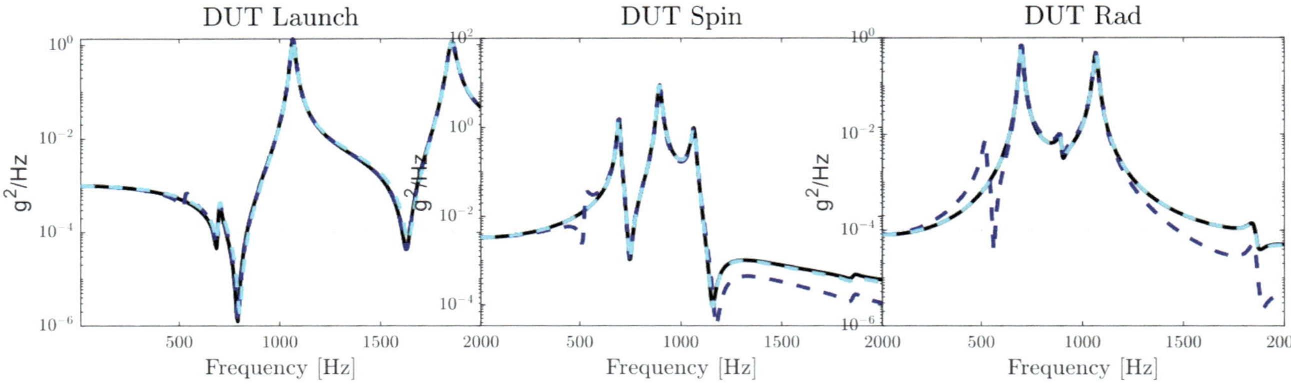

Fig. 15 DUT Environment Reconstruction in two cases. (Dark Blue) Using the shakers in Fig. 13 to control to the 54 next assembly accelerometers in Fig. 8. (Cyan) Using the 10 shakers in Fig. 13 to control to the 5 triaxial DUT accelerometers in Fig. 14. (Black) True Flight Environment

Conclusions

In recent years, methods for accurate vibration qualification have improved significantly, though a major obstacle is that these methods generally require more environment data, which is typically not available. The goal of this paper, therefore, was to understand how accurately component level environments could be estimated given measurements on the operational next assembly only. A MIMO test was simulated showing that DUT response reconstruction is more accurate when environment data is available on the DUT and not just the next assembly. Theoretical conditions for successful environment response reconstruction were presented, and the theory was applied to three case studies.

The first case study showed that error is incurred in response estimation when the amount of next assembly included in the SEREP model does not accurately represent the dynamics of the DUT and next assembly in the flight configuration. The second case study showed that the response reconstruction improved considerably when measurements on the full in flight next assembly were available to use in the test, although noticeable error remained. It was shown that the DUT motion was not fully observable with accelerometers on the next assembly only, and this lack of observability caused error in response reconstruction. The third case study showed that all system modes could be observed if the environment was measured at only a few locations on the DUT. In this case the mode shape matrix was well-conditioned and the environment was estimated accurately.

It is unlikely that one will generally have full-field DUT environment data. This study has shown that it can be accurately estimated, although for the system studied here it was necessary to measure the environment on the next assembly as well as with a few accelerometer channels on the DUT.

Acknowledgments The authors gratefully acknowledge the Department of Energy's Kansas City National Security Campus, operated by Honeywell Federal Manufacturing & Technologies LLC, for funding this work under contract number DE-NA0002839.

References

1. D. Gregory, F. Bitsie, and D. Smallwood, "Comparison of the Response of a Simple Structure to Single Axis and Multiple Axis Random Vibration Inputs," Sandia National Laboratories, Albuquerque, NM, Sandia Report SAND2008-6945C, 2008.
2. R. M. French, R. Handy, and H. L. Cooper, "A Comparison of Simultaneous and Sequential Single-Axis Durability Testing," Experimental Techniques, vol. 30, no. 5, pp. 32–37, Oct. 2006.
3. T. Brown et al., "Multi-axis Vibration Solder Fatigue Analysis Vs. Traditional Vibration Testing," in Proceedings of the 42nd International Modal Analysis Conference, Orlando, FL, Jan. 2024.
4. W. E. Whiteman and M. S. Berman, "Fatigue Failure Results for Multi-Axial Versus Uniaxial Stress Screen Vibration Testing," Shock and Vibration, vol. 9, pp. 319–328, 2002.

5. T. D. Scharton, "Force Limited Vibration Testing Monograph," NASA, RP-1403, 1997.
6. S. Hoffait, F. Marin, D. Simon, B. Peeters, and J. Golinval, "Virtual shaker testing at V2i: measured-based shaker model and industrial test case," in Proceedings of the 2016 International Conference on Noise and Vibration Engineering (ISMA), Leuven, Belgium, Sep. 2016.
7. P. Daborn, "Smarter dynamic testing of critical structures," Ph.D., University of Bristol, 2014. Accessed: Oct. 04, 2022. [Online]. Available: https://ethos.bl.uk/OrderDetails.do?uin=uk.bl.ethos.683391
8. C. Roberts and D. Ewins, "Multi-axis vibration testing of an aerodynamically excited structure," Journal of Vibration and Control, vol. 24, no. 2, pp. 427–437, Jan. 2018, doi: 10.1177/1077546316642064.
9. C. Schumann, M. Allen, M. Tuman, W. DeLima, and E. Dodgen, "Transmission Simulator Based MIMO Response Reconstruction," Exp Tech, vol. 46, no. 2, pp. 287–297, Apr. 2022, doi: 10.1007/s40799-021-00454-4.
10. B. Zwink, "Dynamic Response Matching from Field to Laboratory Replication Methodology," PhD Dissertation, University of Massachusetts Lowell, 2020.
11. J. Reyes-Blanco, "Adjustment of Input Excitation to Account for Fixture-Test Article Dynamic Coupling Effects," PhD Dissertation, University of Massachusetts Lowell, 2017.
12. J. Harvie and R. Mayes, "Quantification of Dynamic Differences Between Boundary Conditions for Environment Specification Improvement," in Shock & Vibration, Aircraft/Aerospace, Energy Harvesting, Acoustics & Optics, Volume 9, in Conference Proceedings of the Society for Experimental Mechanics Series. Cham: Springer International Publishing, 2016, pp. 117–131. doi: 10.1007/978-3-319-30087-0_11.
13. R. Schultz and G. Nelson, "Techniques for Modifying MIMO Random Vibration Specifications," in Sensors and Instrumentation, Aircraft/Aerospace and Dynamic Environments Testing, Volume 7, in Conference Proceedings of the Society for Experimental Mechanics Series. Cham: Springer International Publishing, 2023, pp. 71–83. doi: 10.1007/978-3-031-05415-0_7.
14. R. Barrett, "Techniques for Predicting Localized Vibratory Environments of Rocket Vehicles," National Aeronautics and Space Administration, Huntsville, AL, Technical Note D-1836, Oct. 1963.
15. G. Klein and A. Piersol, "The Development of Vibration Test Specifications for Spacecraft Applications," National Aeronautics and Space Administration, Los Angeles, CA, Contractor Report CR-234, May 1965.
16. J. S. Archer, "Structural Vibration Prediction," National Aeronautics and Space Administration, SP-8050, Jun. 1970.
17. R. Mayes, L. Ankers, and P. Daborn, "Predicting System Response at Unmeasured Locations," Experimental Techniques, vol. 44, Apr. 2020, doi: 10.1007/s40799-020-00366-9.
18. J. O'Callahan, P. Avitabile, and R. Riemer, "System Equivalent Reduction Expansion Process," in Proceedings of the 7th International Modal Analysis Conference, Las Vegas, NV, 1989.
19. M. H. A. Janssens, J. W. Verheij, and T. Loyau, "Experimental example of the pseudo-forces method used in characterisation of a structure-borne sound source," Applied Acoustics, vol. 63, no. 1, pp. 9–34, Jan. 2002, doi: 10.1016/S0003-682X(01)00023-8.
20. Y. Chen, P. Logan, P. Avitabile, and J. Dodson, "Non-Model Based Expansion from Limited Points to an Augmented Set of Points Using Chebyshev Polynomials," Exp Tech, vol. 43, no. 5, pp. 521–543, Oct. 2019, doi: 10.1007/s40799-018-00300-0.
21. M. J. Balas, "Modal Control of Certain Flexible Dynamic Systems," SIAM J. Control Optim., vol. 16, no. 3, pp. 450–462, May 1978, doi: 10.1137/0316030.
22. T. Hall, "Analytically Investigating Impedance-Matching Test Fixtures," in Sensors and Instrumentation, Aircraft/Aerospace, Energy Harvesting & Dynamic Environments Testing, Volume 7, C. Walber, P. Walter, and S. Seidlitz, Eds., in Conference Proceedings of the Society for Experimental Mechanics Series. Cham: Springer International Publishing, 2020, pp. 21–31. doi: 10.1007/978-3-030-12676-6_2.
23. M. Behling, M. Allen, R. Mayes, and W. DeLima, "Understanding Fixture Design in IMMATs Through FEM Simulations," in Proceedings of ISMA 2024, Leuven, Belgium, Sep. 2024.
24. J. P. Salter, "Problem Areas in Dynamic Testing," in Proceedings of the Technical Meeting of the Institute of Environmental Sciences, 1963, pp. 49–54.

Chapter 13
Dynamic Analysis of a Modular Test Stand for Multi-Axis Vibration Testing

H. R. Kramer, J. F. Schultze, S. M. Danforth, and B. P. Mann

Abstract This study investigates a modular test stand, designed to support multi-axis vibration testing, that uses three orthogonal electrodynamic shakers. To capture the effects of shaker-stand interaction, the dynamics of the test stand were analyzed in three unique configurations. The modal characteristics of the test stand in each configuration were extracted via roving hammer impact tests. The approximate transmissibility between various points on the test stands indicated that all three configurations were compliant in the frequency range of interest. Additionally, multiple-input, multiple-output (MIMO) random vibration tests (RVT) were performed on a basic geometry with well-defined dynamics. These MIMO tests analyzed the degree of coupling between the test article and the respective test stand configurations. Examination of the control power spectral densities and inverse frequency response functions revealed that the test stand configuration had no noticeable effect on the performance of the MIMO tests. Furthermore, the resonances of the test article were the main contributor to poor controller performance.

Keywords Test Stand · MAVT · RVT · Modal Analysis · MIMO

Introduction

When attempting to replicate a random vibration environment via ground testing, two common excitation methods are single-axis and multi-axis testing. Multi-axis vibration testing (MAVT) has gained interest in the dynamic environment testing community due to its more realistic environmental replication when compared to traditional single-axis testing. However, MAVT comes with additional design challenges and more complex test setups. When performing single-axis testing, the test article is excited and controlled one axis at a time, often using one large shaker or slip table. Conversely, multi-axis testing aims to excite and control a test article in multiple axis simultaneously, requiring multiple shakers and a test stand. Some work has been done to optimize shaker attachment and response measurement locations for multiple-input, multiple-output (MIMO) vibration tests in [1, 2]. However, every unique test article can result in different optimal shaker attachment points, thereby necessitating a test stand that is modular, easy to assemble, and cost-effective.

Recent advancements in MAVT, such as impedance matched multi-axis testing (IMMAT) [3], have shown that boundary conditions are an important consideration to achieve realistic environmental excitation. Boundary conditions are commonly influenced by test article fixturing and shaker attachment. When a test article is attached to a shaker via a stinger, some shaker-structure interaction will occur. This interaction couples the dynamics of the shaker and stinger to the dynamics of the structure. Shaker-structure coupling can introduce unwanted dynamics to the system, and in turn leads to unrealistic responses. Research has been done to understand this phenomenon in [4, 5, 6, 7, 8] and practical considerations are discussed to mitigate its effects in [9]. However, less published work has been directed toward understanding the dynamics between the shaker and test stand.

Within distributed shaker test setups, methods for positioning shakers include suspending the shakers from or mounting the shakers to a test stand. Suspending the shaker using a bungee reduces the coupling between the shaker and test stand, but

H. R. Kramer · B P Mann
Department of Mechanical Engineering and Materials Science, Pratt School of Engineering, Duke University, NC 27708
e-mail: hunter.kramer@duke.edu; brian.mann@duke.edu

J. F. Schultze · S. M. Danforth
Los Alamos National Laboratory, NM 87545
e-mail: schultze@lanl.gov; sdanforth@lanl.gov

Alexandra Karlicek et al. (eds.), *Dynamic Environments Testing, Vol. 7*, Conference Proceedings of the Society for Experimental Mechanics Series, https://doi.org/10.13052/9788743801528-13

also reduces the force that can be transmitted to the test article. When attempting to replicate more rigorous test specifications, such as flight-relevant environments, the shaker may push itself away from the test article instead of producing the desired environment. Alternatively, the shaker can be rigidly mounted to a test stand. With a rigid structure to push from, the shaker is capable of transmitting a larger force than if it were suspended. However, when the shaker is rigidly attached to the test stand, the shaker and stand become more dynamically coupled. This coupling can introduce unwanted test stand resonances into the MAVT, thus decreasing control quality.

The goal of this study was to measure the effect of test stand dynamics on a sample MIMO vibration test. We first introduce a modular test stand designed to support distributed shaker MAVT with three orthogonal electrodynamic shakers directly mounted to the stand. To understand and possibly mitigate the effects of shaker-stand interaction, we investigated the dynamics of the test stand via modal analysis for the three test stand configurations. Once the dynamics of the test stand and test article were understood, we conducted a MIMO random vibration test for each configuration. The MIMO tests provided a means of assessing the degree of coupling between the test stand and test article.

The remainder of this paper is organized as follows: The three test stand configurations are introduced, followed by an analysis of their dynamics via modal testing and transmissibility considerations. Our MIMO test setup is then described and the results of these tests are discussed. The results are summarized in the discussion/conclusion section, followed by the references and the appendix. The appendix contains the modal frequencies of the three test stand configurations and test article.

Test Stand Configurations

This study examines the dynamics of three unique test configurations. Our first configuration was the original design, referred to as the "Undamped" configuration, and is shown in Figure 1a. We developed our test stand using the Creo Parametric CAD suite. In the CAD design phase, the test stand was iterated upon based on design review feedback from our team. Our design requirements led to a simple, modular, and cost effective test stand. Our finalized design utilized t-slot framing with accompanying hardware that is readily available from hardware suppliers.

Our original design consisted of three compartments, each approximately 1'x1'x3'. The center compartment housed the test article and was spacious enough to mount a 50 lb shaker oriented in the Z-axis (not depicted). The two side compartments, adjacent to the center compartment, provided space for smaller, 7 lb Modal Shop shakers. Our smaller shakers were mounted via their trunion to a rigid aluminum plate. We then mounted the aluminum base plate to the test stand using t-slot hardware. As depicted in Figure 1, the shaker platforms were installed halfway up our vertical beams (1.5'). Our hardware selection facilitated quick reconfiguration, allowing for the small shakers to be moved in three axes. This flexibility was intended to eliminate the need to disassemble or redesign the stand when testing different structures.

To improve upon our initial design, we investigated ways to add stiffness and damping to the test stand. We hypothesized that making the test stand stiffer and more damped would help mitigate the effects of test stand dynamics on the test article. Two subsequent test stand configurations were developed to test this hypothesis. Our first subsequent configuration, referred to as the "Damped" configuration, utilized anti-slip rubber inserts to add damping. We added damping material to one side of every beam in the test stand, however, some locations required extra damping. We evaluated where to add extra damping

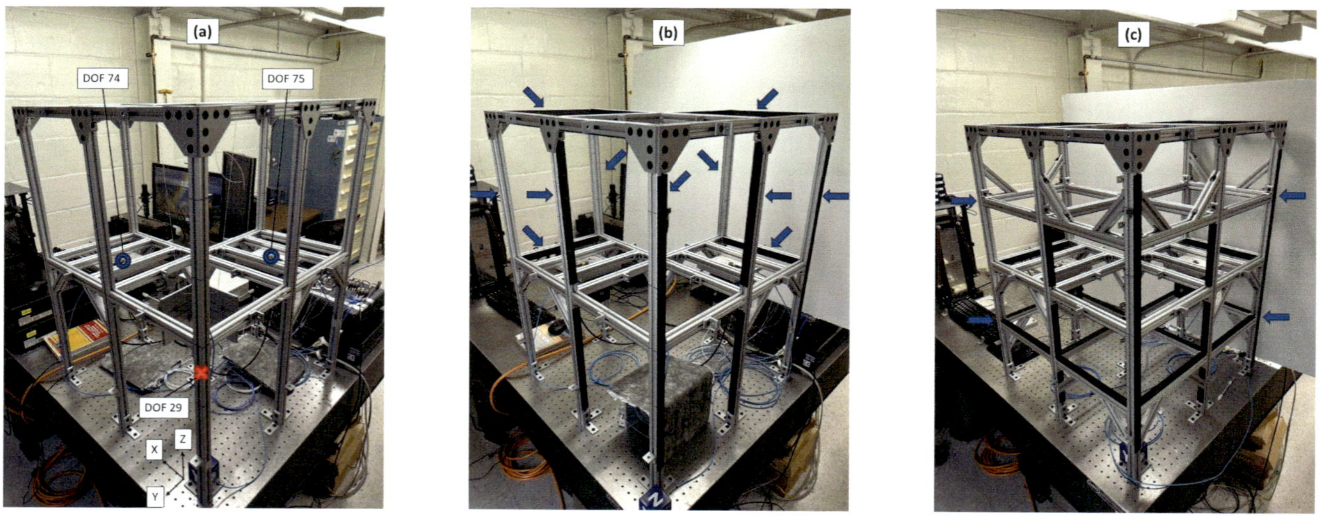

Fig. 1 The three test stand configurations: (a) undamped, (b) damped, (c) stiffened

by impacting the structure with a metal object and listening for an audible ring. The top and middle horizontal platforms were identified via this method to have excessive ringing, and therefore received damping on two sides instead of one. The eight vertical beams and two platforms of our structure where damping material was added are denoted with blue arrows in Figure 1b. Our second subsequent configuration, referred to as the "Stiffened" configuration, aimed to increase the stiffness of the test stand and is shown in Figure 1c. We added horizontal stiffeners and 45° gussets to reduce the distance between supports from 1.5' to 0.75'. As a result, two additional platforms were created and are denoted by blue arrows in Figure 1c. We evaluated the new platforms for excessive ringing and added damping material via the previously described method.

Modal Analysis

To determine the dynamics of the three configurations (undamped, damped, and stiffened) we performed roving hammer impact tests. During each impact test, the small 7 lb shakers were attached to the test stand, but the test article was not. We utilized the Siemens Testlab 2021.2.1 Impact Testing module to estimate frequency response functions (FRFs) for each configuration. These FRFs were then exported to ATA's IMAT 8.1.0 suite to compute the mode shapes via the OPoly application. Each impact test consisted of 78 points excited in three axes, X, Y, and Z. Altogether, we calculated the mode shapes using responses from 234 degrees of freedom (DOFs) for each test stand configuration. To excite the structure in the axial direction of each beam, we installed a 1"x1" L-bracket on the test stand. We then impacted the L-bracket in the axial direction of the beam, allowing for a full modal model to be developed. Between DOFs, we loosened the L-bracket and moved it to the next DOF along the same beam. We recognize that moving the L-bracket between impacts could affect the dynamics of the structure. However, we assumed that the weight of the 1"x1" L-bracket was negligible in comparison to the weight of the test stand.

Our planned MIMO test had a frequency range of 20-2000 Hz. As such, we attempted to obtain the test stand modes up to 2000 Hz for each configuration. Unfortunately, the test stand had many more modes than we initially anticipated. Due to the modal density, we were only able to resolve mode shapes up to 300Hz. In this reduced range, 20-30 modes were extracted, depending on the test stand configuration. To validate our extracted mode shapes, we employed the modal assurance criterion (MAC).

From theory, the modes of a given structure will have unique shapes and corresponding frequencies, except for symmetric geometries which may exhibit repeated roots (such as geometries with square or circular cross sections). The MAC provides an effective means of comparing the similarity of one mode shape to another. The MAC between the mth mode of structure A, $\{\Psi_A\}_m$, and the nth mode of structure B, $\{\Psi_B\}_n$, is calculated as

$$MAC(m,n) = \frac{\left| \{\Psi_A\}_m^T \{\Psi_B\}_n^* \right|^2}{\left(\{\Psi_A\}_m^T \{\Psi_A\}_m^* \right) \left(\{\Psi_B\}_n^T \{\Psi_B\}_n^* \right)} \tag{1}$$

where the superscript T denotes the transpose and the superscript $*$ denotes the complex conjugate. An "ideal" MAC for a single structure with no repeated roots would show all off-diagonal terms to be 0 (all mode shapes are dissimilar to each other) and all diagonal terms equal to 1 (each mode shape is perfectly similar to itself). In this work the MAC scores are shown as percentages, ranging from 0-100 instead of 0-1. It should be noted that for MACs that compare one structure to another, the "ideal" criteria are not the same. A brief review of the MAC and its applications are provided in [10].

A representative sample of MAC scores for each tests stand configuration are shown in Figure 2. All MAC scores shown in Figure 2 correspond to modes under 130 Hz. For the interested reader, all extracted modal frequencies for the three test stand configurations are displayed in Table 4 of the appendix.

A keen eye will discern that our extracted mode shapes were not perfectly orthogonal. Take, for example, the MAC scores for modes 9 and 10 of the stiffened configuration shown in Figure 2c. The MAC score between these two modes indicates that they were 75% similar. Upon inspection of these mode shapes, it was observed that these two modes were similar, but distinct. High MAC scores in the off-diagonal can result from a number of sources. Incorrect mode selection, spatial aliasing, poor signal to noise, or frequency spacing that is too coarse are common sources. In the case of mode 9 and 10 of the stiffened configuration, we noticed that the impact measurements were noisy and that our frequency spacing may have been too coarse. These two factors led to difficulty resolving modes 9 and 10 using using the OPoly application. Despite these challenges, we deemed the MAC plots seen in Figure 2a-c to be acceptable for this study.

The MAC scores in Figure 2 allowed us to analyze how the dynamics of the test stand changed in each configuration. It was observed that the damped configuration had six additional modes in the same frequency range as the undamped configuration. Using the MAC to compare the mode shapes between the undamped and damped configurations, as seen in Figure 2d, these findings can be investigated.

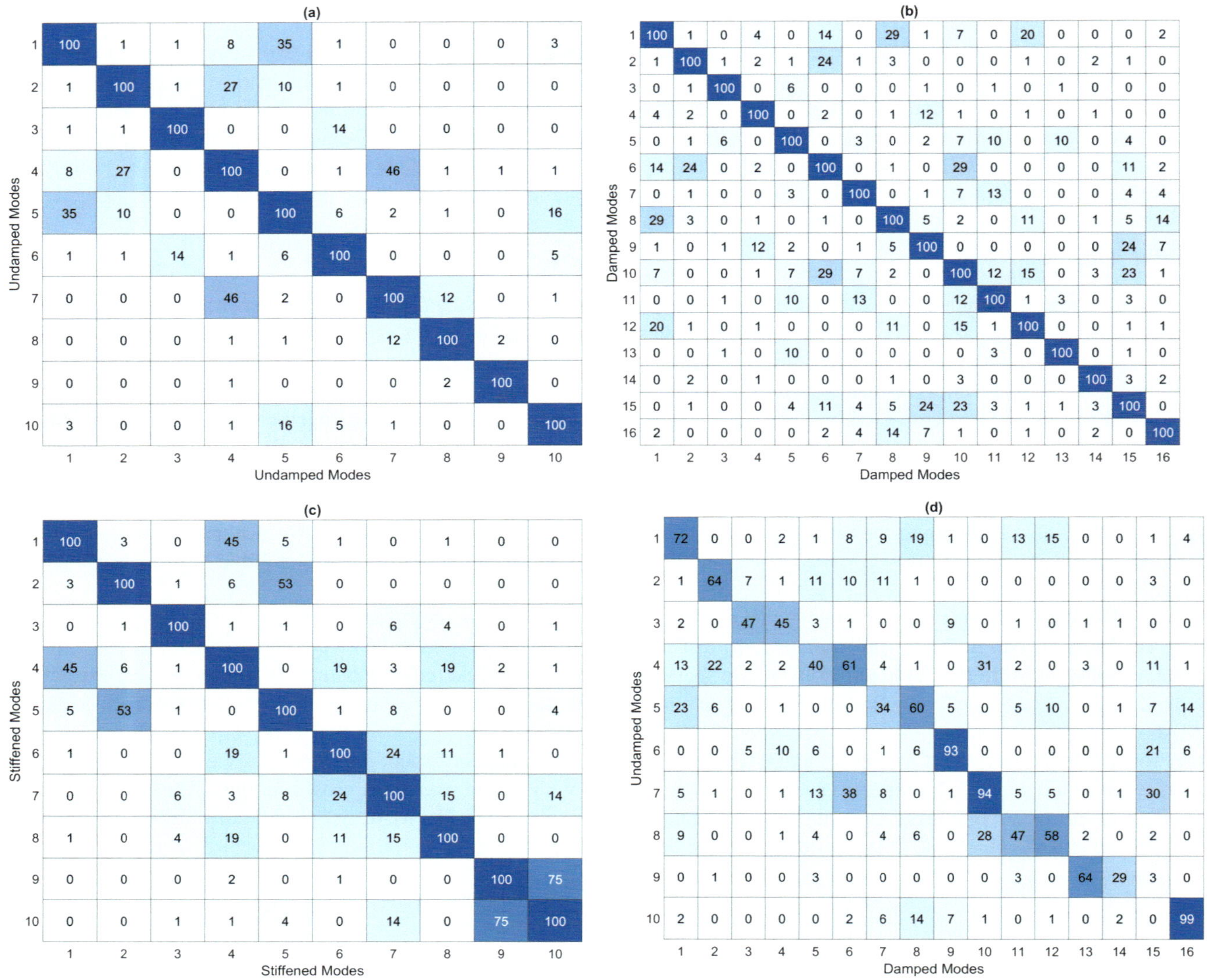

Fig. 2 The modal assurance criterion scores (experimental to experimental) for the three test stand configurations: (a) undamped, (b) damped, (c) stiffened, (d) undamped vs damped

When damping is applied to a system, a single mode can be split into two, lower amplitude modes. Each of the two resultant modes exhibit part of the behavior of the original mode shape. In Figure 2d mode splitting was observed between the undamped and damped configurations. Specifically, modes 3, 4, 5, 8, and 9 of the undamped configuration appear to have split, corresponding to the main diagonal of Figure 2d being stretched along the X-axis. Visual inspection of the extracted mode shapes were consistent with this finding, providing an explanation for five of the six additional modes. The last additional mode, mode 15 of the damped configuration, had very low similarity with any nearby modes. Aside from mode splitting, spontaneous addition of a mode is another potential result of changing the dynamics of a structure (such as adding damping). From this analysis, it can be concluded that the dynamics of the test stand shifted by a measurable amount due to the changes we made.

Transmissibility Analysis

The modal data from our impact tests were useful for identifying dynamic changes between test stand configurations. However, the relative motion between points on the structure was also of interest. One metric that can be useful for understanding relative motion between DOFs is transmissibility. Transmissibility (TR) can be derived using several methods, leading to slightly different interpretations. To avoid ambiguity, a brief review of how we calculated transmissibility is described.

The preferred method for calculating TR is to use spectra obtained from time data. In this method, the TR between response spectra from two DOFs (as opposed to one input DOF and one output DOF for FRFs) is computed using an FRF estimator, such as the H1 method. Doing so reduces the effect of noise on the input (uncorrelated noise will average out). However, we did not have access to the time data from our impact testing. Alternatively, we used a less optimal, FRF-based method. In this method, the TR is approximated using the FRFs for the two response DOFs, and only the resultant magnitude is considered. Suppose an FRF estimate exists between an input at DOF, F, and an output at DOF, X. Using the H1 method, the FRF estimate, H_{XF} is calculated by

$$H_{XF} \approx \frac{G_{XF}}{G_{FF}} \approx \frac{S_X}{S_F} \tag{2}$$

where G_{XF} is the cross power between X and F, G_{FF} is the auto-power of F, S_X is the linear spectra of X, and S_F is the linear spectra of F. Let's assume that an FRF estimate exists for another point, Y, using the same input F. Using the FRF for DOF X, H_{XF}, and the FRF for DOF Y, H_{YF}, the TR between DOF X and Y can be estimated by

$$T_{XY} \approx \frac{H_{XF}}{H_{YF}} \approx \frac{G_{XF}}{G_{FF}} / \frac{G_{YF}}{G_{FF}} \approx \frac{G_{XF}}{G_{YF}} \approx \frac{S_X}{S_Y}. \tag{3}$$

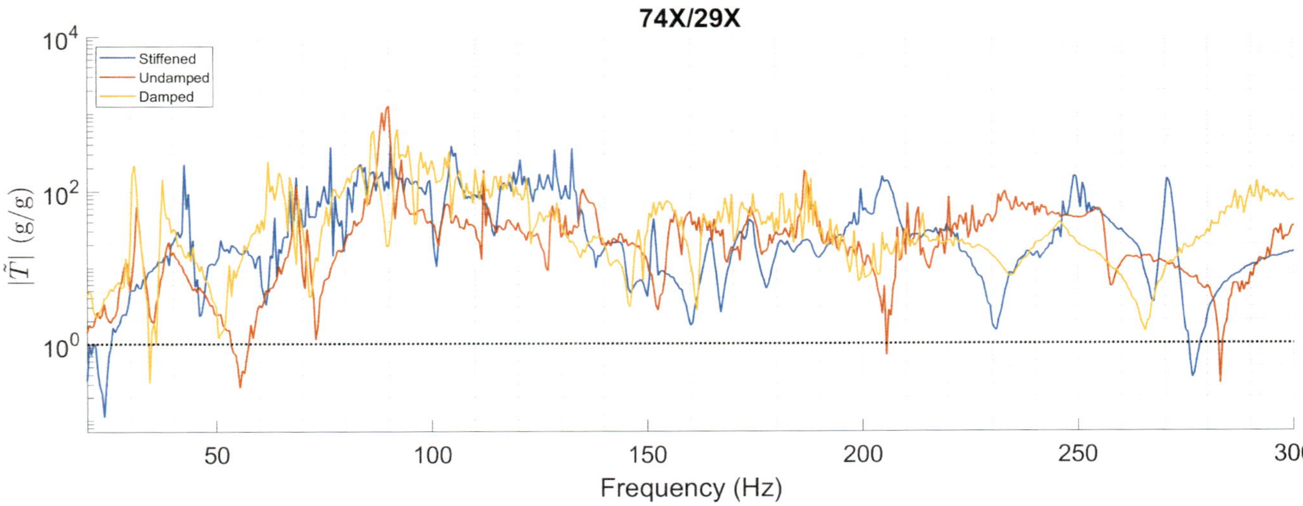

Fig. 3 The approximate transmissibility between the base of the test stand and the shaker attachment point for all three test stand configurations, with a horizontal dotted line added at 10^0 for clarity

Equation 3 shows that the contribution from the input F cancels out and thus leaves the TR between DOF X and Y. In theory, the noise from the G_{FF} cancels out of this estimate. However, in practice, there is always some residual error from the noise. Additionally, this method fails when the applied force approaches zero, indicating a force dropout in the data. With these limitations in mind, the approximate TRs between the test stand configurations were analyzed.

Ideally, shakers rigidly attached to a test stand would allow for the transfer of force without introducing unwanted dynamics. To estimate the deviation of our test stands from this idealization, we examined the magnitude of the approximate TR between the base of the test stand, and the shaker attachment point. For a perfectly rigid test stand, the magnitude of the TR between the base and the shaker equals 1, corresponding to zero relative motion between them. Figure 3 shows the approximate TR between a sample shaker mounting DOF, 74X, and a sample base point of the test stand, 29X. Both locations are denoted in Figure 1a. Note that DOF 29X was located behind the vertical beam in the foreground of Figure 1a and was indeed a foot of the test stand.

The TR depicted in Figure 3 indicates that our test stand was far from rigid. In fact, in the range of our modal analysis, relative motion was observed on the order of 10^3 between the base of the test stand and where the shaker was mounted. Relative motion of this magnitude indicates that our test stand was very compliant in this range. This finding indicates that for all three test stand configurations, the dynamics of the test stand affected the shakers, and may have affected the test article.

In addition to the approximate TR between the shaker locations and the base of the test stand, we also investigated the relative motion between the two shaker locations. Ideally, the magnitude of the TR between these two DOFs would be close to 1, indicating that the shakers will not move relative to each other. Relative motion between the shakers can induce a moment on the test article or reduce the force transmitted through the stingers.

Figure 4 shows the approximate TR between the two shaker locations. We chose DOF 74X and DOF 75Y for this analysis because they were the shaker drive DOFs and are denoted in Figure 1a. High TR between these two DOFs indicates that the shakers, in the direction of the applied force, were moving relative to each other. Figure 4 shows that the relative motion between shakers was much less severe than the motion between the shakers and the base. However, there was still relative motion on the order of 10^2 between them in the range of 100-150 Hz. Adding stiffness to the test stand positively affected the TR in both of our comparisons, pushing the TR curves down toward 1. Therefore, we predicted that the stiffened configuration would have better control performance in the MIMO test.

Our analysis of the mode shapes and approximate TRs of our test stand indicated that dynamic interaction between our test stand and shakers was likely. To ascertain the degree of interaction, we performed MIMO vibration tests in each test stand configuration. The following section introduces our test article and details how we performed our MIMO tests.

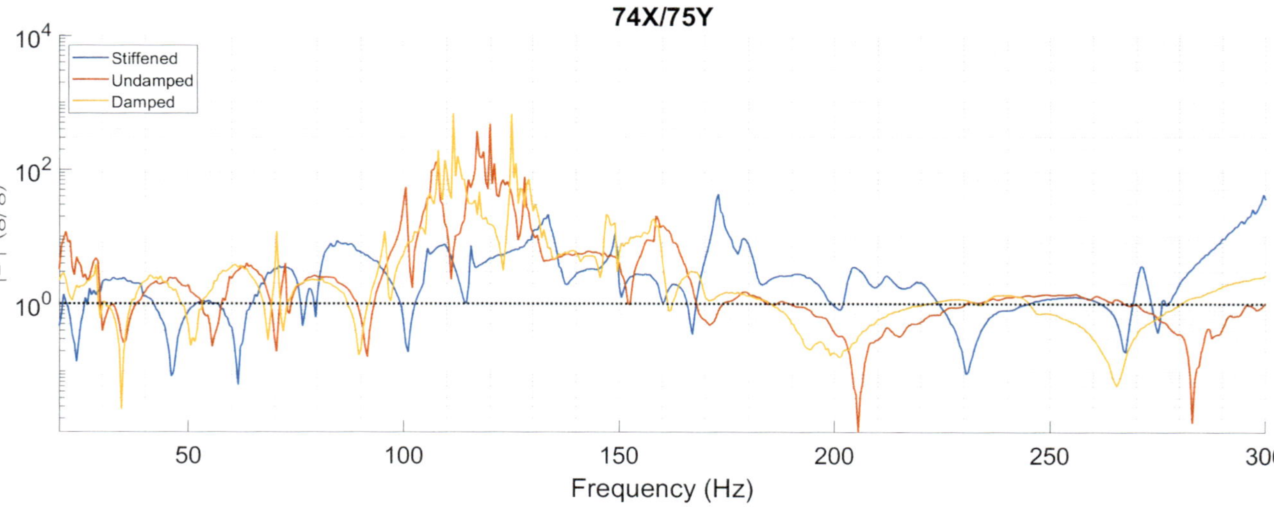

Fig. 4 The approximate transmissibility between the two shaker attachment points for all three test stand configurations, with a horizontal dotted line added at 10^0 for clarity

MIMO Test Setup

An appropriate test article is required to perform a MIMO test. We chose our test article based on two design requirements. The test article needed to have a simple geometry and well-defined dynamics. With a plethora of commercially available options, we chose an aluminum angle and iterated on the dimensions to ensure the test article had interesting dynamics in our frequency range of interest. From a finite element analysis (FEA), eight modes between 20-2000 Hz were expected. We confirmed these expectations through impact modal testing, where the test article was suspended by a fishing line to simulate a free-free boundary condition. The modal frequencies extracted from our impact test were within 5% of our FEA modes. We also visually compared the corresponding mode shapes and observed good agreement. The dimensions of the test article were 6"x6"x6", with a thickness of 1/4". We applied a grid of test points to the test article at a 2" spacing, except at the center line, where we applied a 1" spacing. The location of these grid points were informed by our FEA, and led to well-resolved mode shapes (as evidenced by our MAC in Figure 5c).

The test article, referred to as the Aluminum Angle (AL angle), was easily procured, with a shape that was conducive to attaching two orthogonal shakers (one on each wing). Figure 5 shows the AL angle in several orientations and the computed MAC from our modal analysis. The modal frequencies for the AL angle are listed in Table 4 of the appendix. The lowest mode occurred around 320 Hz. Since the range of our modal test on the test stands was limited to 300Hz, the dynamic effects from the test stand and the test article could be distinguished. We also note that the modal damping calculated for each AL angle mode was on the order of 0.01. Test articles with low damping such as our AL angle are challenging to control.

The goal of our study was to understand how the dynamics of our test stand affected MIMO testing. We developed a sample test specification with sufficient spectral energy to excite the modes of the test article. The control range for our

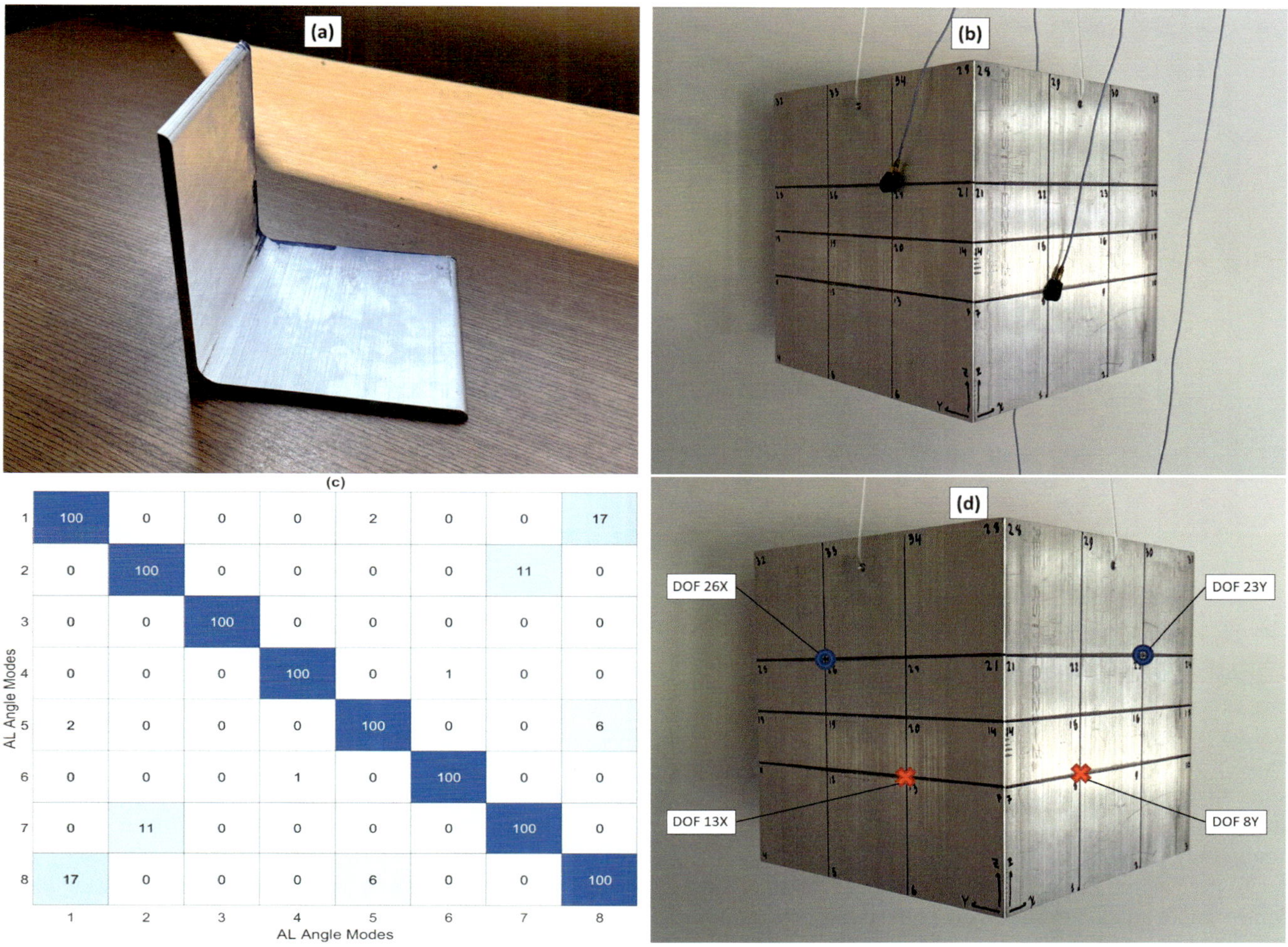

Fig. 5 The test article used in our MIMO tests: (a) AL Angle, (b) AL angle in the modal test configuration, (c) AL angle modal assurance criterion scores (experimental to experimental), (d) the AL angle in the MIMO test configuration

MIMO test was 20-2000 Hz. In the Testlab 2021.2.1 MIMO Random module, the lower end of the control range may not be a whole number (depending on the desired frequency resolution), leading us to develop a specification that started just below 20 Hz. Table 1 shows the breakpoints chosen for our test specification.

Table 1 The test specification breakpoints used for both control DOFs in our MIMO tests

Frequency (Hz)	**Magnitude** (g^2/Hz)
18	$1 \cdot 10^{-4}$
50	$2 \cdot 10^{-4}$
1000	$2 \cdot 10^{-4}$
2000	$1 \cdot 10^{-4}$

Our MIMO test utilized two shakers oriented in the X and Y directions as shown in Figure 6. We attached each shaker to the test article using a nylon stinger with an aluminum sheath. The stinger was then screwed into a force transducer, which we rigidly attached to the test article using super glue. Directly opposite the force transducer, control accelerometers were instrumented on the test article. Our choice of shaker attachment locations on the test article was informed by the experimental modal test. Locations lying on nodes of the modes were excluded to ensure adequate excitation of all the modes of our test article. Using engineering judgement, DOF 26X and 23Y were selected as the shaker attachment points and are denoted with blue circles in Figure 5d. Two additional response locations were instrumented at DOF 13X

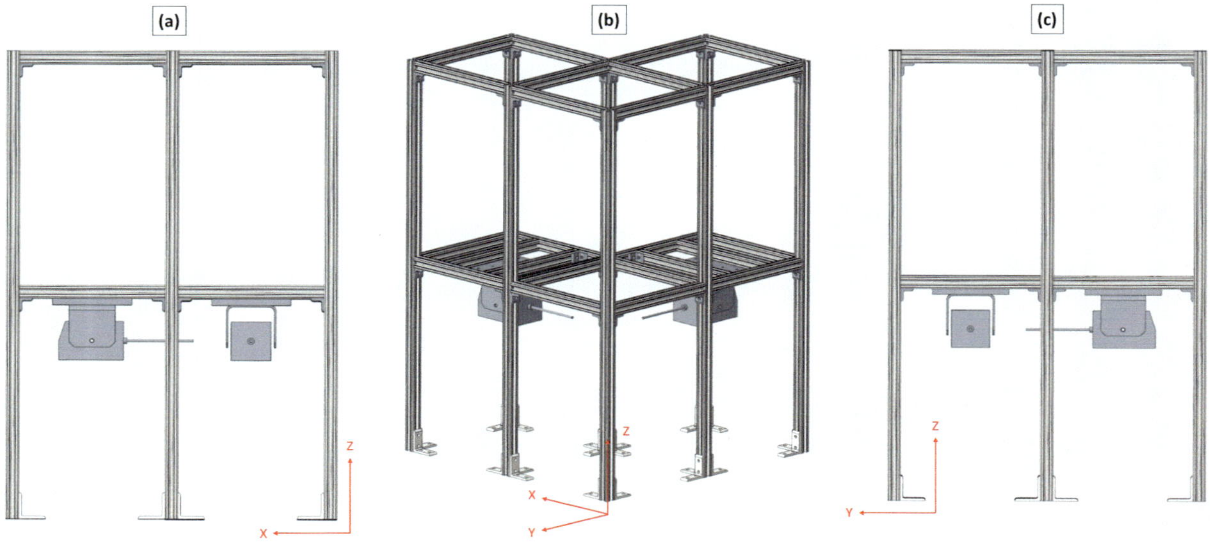

Fig. 6 The CAD model for the undamped test stand configuration with the 45° gussets removed: (a) left view, (b) isometric view, (c) right view

and 8Y to analyze the behavior of the test article at uncontrolled locations. The two response locations are denoted with red X's in Figure 5d. The triaxial accelerometers used during our modal test on the test stand (DOFs 74 and 75) were retained for our MIMO tests. Using these accelerometers, we were able to measure the response of the test stand, if any, at the shaker attachment points. Figure 1a shows the test article set up for the MIMO test in the undamped test stand configuration.

In the Testlab 2021.2.1 MIMO Random module, the user has the option to choose between control modes. In the PSD control mode, the controller ignores the cross power spectral density (CPSD) terms and only controls to the specified PSDs. An alternative control mode is "Matrix" control. In this mode, the controller attempts to control the system to specified CPSDs and PSDs. Our test specification did not include CPSDs, so PSD control was selected.

The sequence of tests followed a consistent pattern: The test stand configuration was assembled, a modal test was performed on the test stand, the test article was attached to the test stand, and then MIMO tested. Between assembly of the test stands, the test article was dismounted and set aside, allowing for easier assembly of the next test stand configuration. In an effort to keep the boundary conditions of the test article consistent with the modal test, we suspended the test article with fishing line for the MIMO tests. Wax was used to mount sensors to the test article, and accelerometers disturbed by removing and re-mounting the the test article where reseated as needed. The wax used in this study was sufficient up to 2000 Hz, and was chosen over a stronger adhesive to extend the life of the accelerometers.

MIMO Test Results

For the undamped and damped test configurations, the MIMO tests were performed successfully. However, when we performed the MIMO test in the stiffened configuration, an audible rattling could be heard around 1200 Hz. In addition to the rattling, we observed excessive twisting and displacement of the test article. We made several attempts to eliminate the twisting and displacement but were ultimately unsuccessful. Because there were no safety issues with this rattling, we adjusted the percentage of frequency lines that could violate the alarm (± 3 dB) and abort (± 6 dB) limits to allow the stiffened MIMO test to run. As a result, the full test was completed but exhibited poor low-frequency control. In searching for methods to reduce rattling, we truncated the specification to 20-1200 Hz and re-ran the stiffened MIMO test. Truncating the higher frequencies resulted in adequate low-frequency control. The results from the truncated MIMO test will be referred to as"Stiffened/Truncated." However, the results from the stiffened 20-2000 Hz MIMO test were retained for completeness.

Control Comparison

It is common practice when comparing random vibration tests to plot the control PSDs for each test against one another. Figure 7 plots the control PSD curves for all four MIMO test cases: undamped, damped, stiffened, and stiffened/truncated. Along with the PSD curves, the test specification, alarm (±3dB), and abort (±6dB) lines are plotted. The first row of Figure 7 compares each of the full bandwidth tests against each other. The second row compares only the stiffened test to the stiffened/truncated test. Each column of Figure 7 plots the control curves for a specific control DOF.

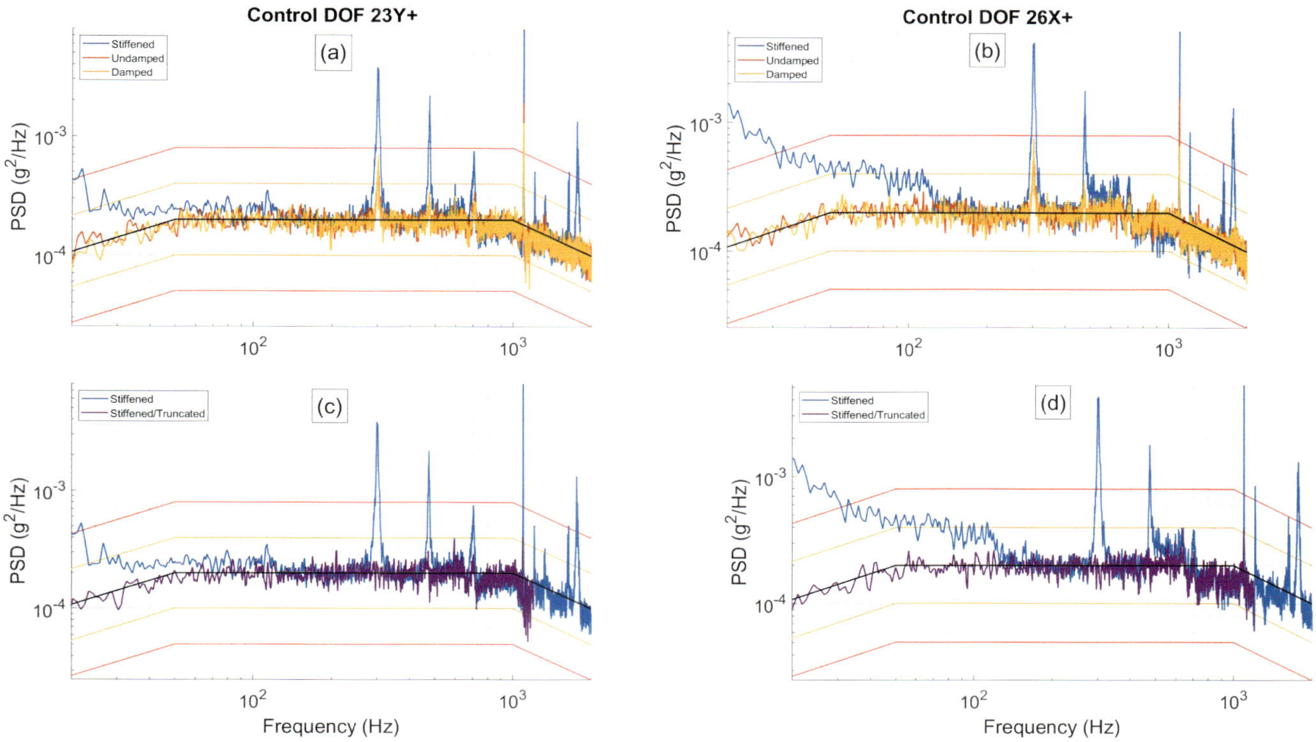

Fig. 7 The control PSD curves from the various MIMO tests. (a): undamped vs damped vs stiffened control at DOF 23Y+, (b): undamped vs damped vs stiffened control at DOF 26X+, (c): stiffened vs stiffened/truncated control at DOF 23Y+, (d): stiffened vs stiffened/truncated control at DOF 26X+

In Figure 7, the control PSDs for the undamped and damped configurations are nearly indistinguishable. One small difference is that the damped configuration displayed a small decrease in the peak seen around 1100 Hz. The stiffened control curve shows the greatest deviation from the test specification, as was expected with the issues we encountered.

The MIMO controller did not appear to have any issues controlling the test article between 20-300 Hz (excluding the stiffened configuration). Despite the multitude of test stand modes in this range, the control PSDs were well within the alarm limits. While the test stand modes seemed to have little impact on the MIMO control, the control was poor near the modes of the test article. The peaks in the PSD curves violated the abort limits at the following frequencies: 300 Hz, 475 Hz, 1096 Hz, 1208 Hz, 1624 Hz, and 1771 Hz. These peaks did not line up exactly with these test article modes, but were reasonably close (with the exception of 1208 Hz, where the rattle was observed). It is not uncommon for shaker-structure interaction to introduce slight dynamic changes to the test article. Therefore, it is reasonable to assume that some of the free-free modes of the AL angle may have shifted. Interestingly, the test article modes did not seem to pose a challenge for the controller in the stiffened/truncated case. The test stand did not change between the stiffened and stiffened/truncated MIMO tests, so the difference in performance was due to changes in the test article or control setup. One caveat to the stiffened/truncated MIMO test was the limited frequency range. Limiting the number of frequencies the controller must manage at any given time creates an easier control problem to solve.

It is difficult to differentiate the performance of similar control curves without some sort of metric. While limited, some standards do exist for comparing the performance of MIMO tests. One such standard found in MIL-STD-810H [11] is the acceleration root mean squared (gRMS) error. The gRMS error is used to determine the percent difference between the gRMS of the test specification, and the achieved control. The gRMS error was calculated using

$$gRMS_{Error} = \frac{|gRMS_{Spec} - gRMS_{Control}|}{gRMS_{Spec}} \cdot 100 \tag{4}$$

where $gRMS_{Spec}$ is the gRMS of the test specification, and $gRMS_{Control}$ is the gRMS of the achieved control. The lower the gRMS error score, the closer the control was to achieving the test specification. Table 2 displays the % gRMS error for each MIMO test at each control DOF.

The data in the far right column of Table 2 has been shaded based on performance of the respective test. Cells highlighted in green had better performance, while cells highlighted red had poorer performance. Unsurprisingly, the stiffened MIMO test overshoots the test specification by a wide margin. Analyzing the remaining MIMO test, it was observed that the damped MIMO test had the best performance. However, the % gRMS error between the the undamped, damped, and stiffened/truncated MIMO test only vary by about 3%. The threshold for gRMS error from MIL-STD-810 is ±10%. The undamped, damped, and stiffened/truncated MIMO tests were well below this threshold. This finding was surprising considering the dynamic differences between the test stand configurations.

The next metric applied to the data was the abort/alarm violation error. While not found in a published standard, this metric is commonly used in the Department of Energy (DOE) complex. When performing a MIMO test in Testlab 2021.2.1, abort and alarm lines are plotted with the test specification and measured control. The abort and alarm lines correspond to ±6dB and ±3dB from the test specification respectively. Applying a threshold for abort/alarm line violations can serve as a quick check of test quality before concluding a vibration test. This metric is calculated by summing all control frequency lines outside of the abort/alarm limits, and normalizing by the total number of lines. Table 3 summarizes the abort/alarm line performance of our MIMO tests.

Table 3 is shaded in a similar fashion to Table 2, where green cells represent better performance, and red cells represent poorer performance. The stiffened MIMO test again performed the worst by a large margin. The remaining tests, however, do not show a large difference in control performance. This metric was more rigorous for the stiffened/truncated MIMO test because the limited frequency range reduces the number of spectral lines for normalization. Therefore, it takes fewer violations to get the same % error as the other tests with full ranges. Despite this, the differences in abort violations between the undamped, damped, and stiffened/truncated MIMO tests were less than 1%. Comparing the alarm violations for these tests, the error was larger, but the maximum difference between configurations was still only around 1%. The threshold for

Table 2 The gRMS % Error scores for the four MIMO test at each control DOF

		gRMS		
Configuration	**Control DOF**	**Control**	**Test Specification**	**% Error**
Undamped	23	0.597	0.571	4.6
	26	0.586	0.571	2.6
Damped	23	0.559	0.571	2.1
	26	0.560	0.571	1.9
Stiffened	23	0.651	0.571	14.0
	26	0.662	0.571	15.9
Stiffened/Truncated	23	0.466	0.473	1.5
	26	0.457	0.473	3.4

Table 3 The % abort/alarm line violation scores for the four MIMO test at each control DOF

		% Violation	
Configuration	**Control DOF**	**Alarm**	**Abort**
Undamped	23	0.63	0.20
	26	0.47	0.12
Damped	23	0.71	0.12
	26	0.39	0.00
Stiffened	23	7.57	2.88
	26	10.37	3.75
Stiffened/Truncated	23	1.72	0.00
	26	0.60	0.00

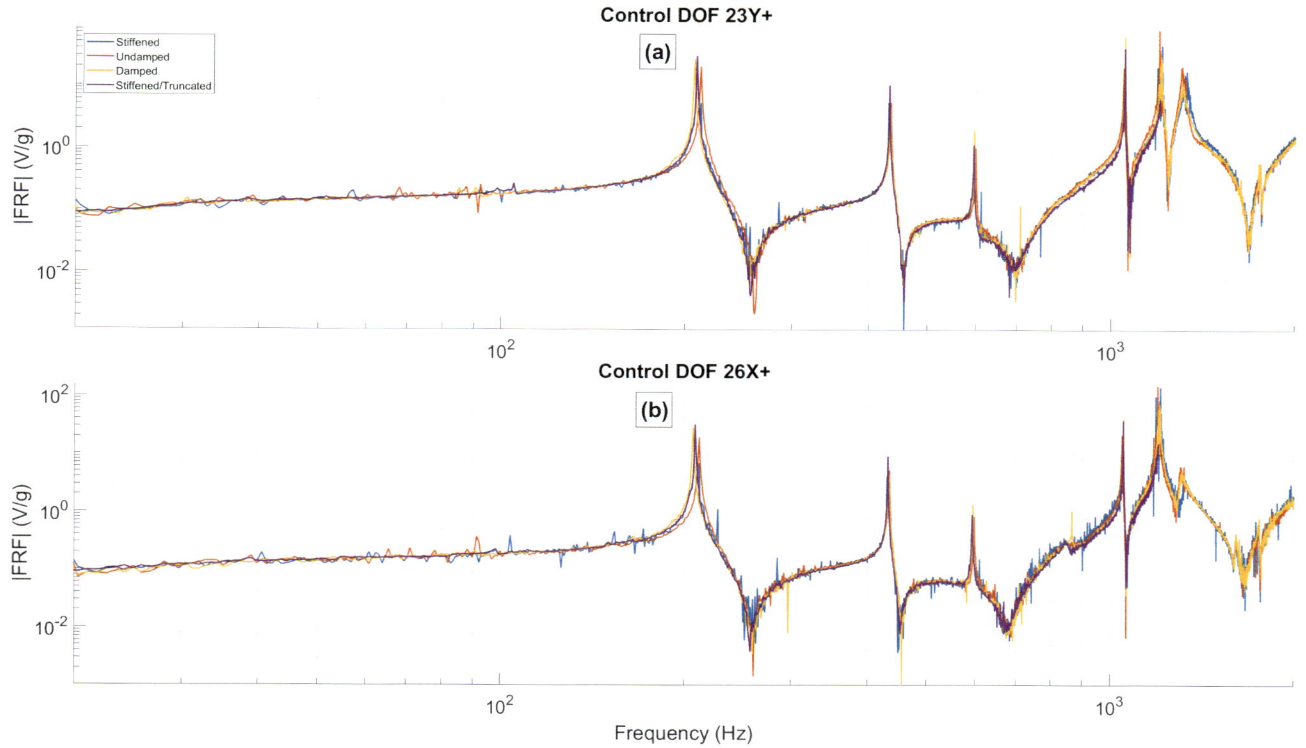

Fig. 8 The inverse FRFs reported by Testlab 2021.2.1 for the four MIMO tests: (a) inverse FRFs for control DOF 23Y+, (b) inverse FRFs for control DOF 26X+

alarm/abort line violations is 5% and 2% respectively. The undamped, damped, and stiffened/truncated MIMO tests were again well below these thresholds. The results from this metric agree with our findings from the previous metric, suggesting little impact from the test stand on the MIMO tests.

Inverse FRF Comparison

Our previous results showed that the dynamics of the test stand in the range of identified modes (20-300 Hz) did not appear to influence the control performance of the MIMO tests. However, the question remained whether the controller overcame the test stand dynamics, or was not affected by them at all. To understand the impact the test stand dynamics on the controller, we analyzed the inverse FRFs between the control DOFs and input voltage from our MIMO tests. These FRFs inform the controller's decisions concerning how much voltage to apply to get a desired acceleration response. Resonances from the test stand would appear in the inverse FRFs as peaks and valleys if the controller were making an effort to overcome them. Figure 8 plots the inverse FRFs reported by the Testlab 2021.2.1 MIMO random module for each of the MIMO test cases.

The inverse FRFs from Figure 8 were relatively flat in the range of 20-200 Hz, where many of our identified test stand modes lie (16-25 modes depending on the test stand configuration). Since there were no peaks or valleys, one can conclude that the test stand resonances had no noticeable impact on the controller in this range. It is observed that there was a resonance and anti-resonance in the inverse FRF between 200-300Hz, and the test stand had resonances in this range. However, if this behavior were caused by variations in the test stand, it would be expected that the inverse FRF would vary between configurations. Instead, the system dynamics had almost *no* change from test to test. The inverse system FRFs were nearly identical regardless of test stand configuration. This implies that the behavior of the test article, as a result of the shaker input, was independent of the test stand dynamics.

The inverse FRFs, see Figure 8b, were observed to have more distortion than the curves in Figure 8a. The audible rattling describe in earlier sections emanated from the shaker located at DOF 26X+. This rattling increased the level of noise in the measured data, and was likely the cause of the added distortion. Because the distortion appears in all of the curves, the rattling may have been a issue in all of the MIMO tests, progressively getting worse until the controller could no longer handle it in the stiffened MIMO test. Limiting the frequency range of the stiffened/truncated MIMO test cut off the rattle frequency, and allowed the controller to replicate the specification with greater accuracy.

Discussion and Conclusion

Performing a modal test on a test stand can be a useful tool for understanding at what frequencies test stand resonances occur. Identifying these resonances can give the test engineer confidence in determining whether undesirable control performance was caused by the test article or the test stand. From the high modal density in the low frequency range to the high compliance from the transmissibility plots, our test stand exhibited many undesirable behaviors. Based on these observations, we expected that the resonances of the test stand would negatively impact our MIMO control. However, this expectation was not substantiated by the experimental evidence. Instead, our analyses indicated that the dynamics of the test stand had little discernible impact on our MIMO tests in this case. Furthermore, the lightly damped resonances of the test article were the single biggest contributor to poor MIMO control.

This study analyzed a modular test stand for random MAVT. Three test configurations were analyzed, with each intended to be an improvement on the last. We began by performing impact hammer tests on each configuration and extracting the mode shapes. From comparison of the mode shapes via the MAC, we concluded that five of the modes from the undamped configuration were split in the damped configuration. The TR was approximated for each test configuration to capture the rigidity of the configurations. The TR plots indicated that adding stiffening improved the rigidity of the test stand, but did not eliminate compliance. The test article used in MIMO testing was also impact tested and its mode shapes extracted. The test article had eight lightly damped modes in our frequency range of interest, i.e. the frequencies 20-2000 Hz.

We then compared the MIMO test performance of the various test configurations. The stiffened MIMO test performed poorly, likely due to unwanted rattling and displacement of the test article. The other three MIMO tests satisfied our success criteria. The inverse FRFs from our MIMO tests indicated that there was little dynamic change to the shaker-test article dynamics between the test stand configurations. We therefore concluded that the dynamics of our test stand had little to no effect on the control effort in the MIMO tests. Additionally, the dynamics of the test article were the greatest contributing factor in test specification adherence.

We do not claim that these findings will be the case for all test stands used in random MAVT. Other important considerations include the weight of the test article, the size of the shakers, characteristics of the test specification, and the geometry of the test stand. Future work is needed to understand how these factors may impact MIMO test performance and shaker-stand coupling.

Acknowledgments This research was supported by the Delivery Environments program under the Office of Engineering and Technology Maturation at Los Alamos National Laboratory (LA-UR-24-30436).

References

1. Rohe, D.P., Nelson, G.D., and Schultz, R.A. "Strategies for shaker placement for impedance-matched multi-axis testing". In *Sensors and Instrumentation, Aircraft/Aerospace, Energy Harvesting & Dynamic Environments Testing, Volume 7: Proceedings of the 37th IMAC, A Conference and Exposition on Structural Dynamics 2019*, pages 195–212. Springer (2020)
2. Mayes, R., Ankers, L., Daborn, P., Moulder, T., and Ind, P. "Optimization of shaker locations for multiple shaker environmental testing". *Experimental Techniques*, 44:283–297 (2020)
3. Daborn, P.M., Roberts, C., Ewins, D., and Ind, P. "Next-generation random vibration tests". In *Topics in Modal Analysis II, Volume 8: Proceedings of the 32nd IMAC, A Conference and Exposition on Structural Dynamics, 2014*, pages 397–410. Springer (2014)
4. Avitabile, P. "Why you can't ignore those vibration fixture resonances". *Sound and Vibration*, 33:20–27 (1999)
5. Trubert, M. "Structural and electromechanical interaction in the multiple-exciter technique for random vibration testing". *The Journal of the Acoustical Society of America*, 41(5):1185–1192 (1967)
6. Peres, M., Bono, R., and Avitabile, P. "Effects of shaker, stinger and transducer mounting on measured frequency response functions". *The 25th Intenational Conferenece on Noise and Vibration Engineering ISMA-USD*, pages 1475–1490 (2012)
7. Varoto, P.S. and de Oliveira, L.P. "Interaction between a vibration exciter and the structure under test". *Sound and vibration*, 36(10):20–26 (2002)
8. Siswanto, W.A., Ibrahim, M.N., Madlan, M.A., and Mohamad, S.M. "Shaker table design for electronic device vibration test system". *International Journal of Engineering and Technology*, 3(6):663 (2011)
9. Peres, M.A., Bono, R.W., and Brown, D.L. "Practical aspects of shaker measurements for modal testing". In *Proceedings of the ISMA*, volume 1105, pages 2539–2550 (2010)
10. Pastor, M., Binda, M., and Harčarik, T. "Modal assurance criterion". *Procedia Engineering*, 48:543–548 (2012) Modelling of Mechanical and Mechatronics Systems.
11. U.S. Dept. of Defence, Washington, DC, USA. *MIL-STD-810H* (2019)

Appendix

Table 4 The modal frequencies for the three test stand configurations and test article

	Frequency (Hz)			
Mode Number	**Undamped**	**Damped**	**Stiffened**	**AL Angle**
1	29.4	28.7	38.8	323.8
2	32.1	30.8	39.2	501.7
3	39.2	36.0	51.4	679.1
4	68.8	37.4	75.5	1100.4
5	70.9	66.4	79.5	1643.9
6	88.5	67.0	99.4	1694.6
7	93.0	68.2	104.8	1772.0
8	101.9	68.9	105.1	1909.4
9	111.2	86.2	114.7	-
10	126.9	91.5	115.2	-
11	137.8	96.1	144.4	-
12	151.4	97.0	151.6	-
13	154.6	104.4	165.0	-
14	157.9	105.0	171.6	-
15	168.9	110.8	180.8	-
16	175.1	123.0	187.5	-
17	185.3	140.5	201.6	-
18	209.5	146.6	205.9	-
19	255.3	148.9	225.2	-
20	-	161.3	237.9	-
21	-	162.7	264.6	-
22	-	164.2	271.3	-
23	-	168.9	276.1	-
24	-	171.2	279.5	-
25	-	194.7	-	-
26	-	222.2	-	-
27	-	232.3	-	-
28	-	245.3	-	-

Chapter 14
Evaluating the Effect of Shaker Placement Optimization Priorities on Multi-Axis Test Results

Risto Djishev, Kieran Elrod, Connor Tasik, Jim DeClerck, Brittany Ouellette, John Schultze, and Shannon Danforth

Abstract Multi-axis vibration testing is capable of emulating field environments more accurately than traditional single-axis testing. However, the presence of multiple shakers introduces challenges in test setup and execution. Optimization algorithms can help researchers select the location and direction of each shaker in a multi-axis test by minimizing an objective function subject to constraints. One example is to select shaker locations that minimize required shaker force (objective) while matching a power spectral density specification at control degrees of freedom (constraint). The solution given by the algorithm is affected by many variables, including the choice of cost and constraints, metrics used for formulating the cost and constraints, and model fidelity and calibration. When these optimized shaker configurations are implemented in a laboratory setup, differences in boundary conditions, shaker and sensor attachment method, and shaker control are likely to affect how the optimization criteria is reflected in the test results. This case study investigates the challenges and benefits of optimizing shaker location by determining shaker placements from an optimization problem, implementing this suggested shaker configuration in a laboratory test, and comparing results to those from an "engineering judgment" testing configuration.

Keywords Shaker placement optimization · MIMO testing · Vibration control · Finite element modeling · Model updating

Introduction

Environmental testing, or the practice of recreating field conditions in a lab setting for qualification of devices prior to operation, is a widely used process across many industries to ensure successful deployment of products or improve the effectiveness of design iterations. Environmental testing in laboratory settings increases safety and cost effectiveness compared to field testing. In military applications, environmental testing is a legal requirement for new products prior to mass production [1].

This study focuses on random vibration testing, which aims to replicate the acceleration power spectra responses that a device would experience in its operational environment. In this field, testing has traditionally been conducted using

Risto Djishev
Department of Mechanical Engineering, Rice University, Houston, TX 77005
e-mail: rd40@rice.edu

Kieran Elrod
Department of Structural Engineering, University of California, San Diego, La Jolla, CA 92093
e-mail: kieranelrod@gmail.com

Connor Tasik
Department of Mathematics, Rose Hulman Institute of Technology, Terre Haute, IN 47803
e-mail: tasikca@rose-hulman.edu

Jim DeClerck
Department of Mechanical Engineering, Michigan Technological University, Houghton, MI 49931
e-mail: jdeclerck@mtu.edu

Brittany Ouellette · John Schultze · Shannon Danforth
Los Alamos National Laboratory, Los Alamos, NM 87545
e-mail: britt@lanl.gov; schultze@lanl.gov; sdanforth@lanl.gov

Alexandra Karlicek et al. (eds.), *Dynamic Environments Testing, Vol. 7*, Conference Proceedings of the Society for Experimental Mechanics Series, https://doi.org/10.13052/9788743801528-14

single-axis electrodynamic shakers. Single-axis testing typically involves testing X,Y, and Z specifications sequentially, then combining results from each axis to form a global error metric [2]. Although single-axis testing is commonly applied, it is known to have several issues. Cross-axis (i.e., out-of-axis) responses are a common problem with single-axis testing. These responses are difficult to incorporate into the global error metric and can negatively affect the in-axis response data. Similarly, the practice of securing the device under test to a shaker plate often means that the real boundary conditions of the structure are not well represented [3]. Therefore, an accurately described input vibration spectra may not recreate a realistic response because the system impedance is not representative of the field impedance. Because of the aforementioned issues with single-axis tests replicating the real environment, it is common for over-testing to lead to unnecessary damage in some axes/responses while paradoxically under-testing other axes/responses [3].

To address these problems with single-axis testing, multi-input, multi-output (MIMO) random vibration testing has become more popular for recreating vibration environments. Utilizing multiple shakers rather than a monolithic shaker base can allow for a more accurate recreation of vibration sources, reduce damage associated with over-testing by reducing the overall force needed to excite critical modes, and capture multi-axis response [3]. However, the placement of input sources for these tests is often based on the judgment of individual engineers, varying the effectiveness of MIMO shaker testing from practitioner to practitioner. To address this shortcoming, researchers have begun to implement optimization approaches for selecting shaker locations and directions in MIMO tests [4] [5].

Optimization algorithms for shaker placement often rely on estimates of the system's frequency response functions (FRFs) from either a finite element analysis (FEA) model or experimental modal analysis. The optimization uses these FRFs to predict the response of a system to shaker inputs at particular degrees of freedom (DOFs). Synthesizing FRFs from an FEA model provides the advantages of obtaining FRFs at a finer resolution than is possible with experimental modal analysis. Furthermore, in a pre-test stage where access to the physical system is limited, experimental modal analysis may not be possible. However, FEA models represent idealized versions of the system and do not perfectly represent the system's material properties or boundary conditions. If the FEA model lacks enough similarity to the physical system in its tested state, then the cost and constraints implemented in the optimization algorithm may not be reflected in the hardware test results.

This paper investigates the challenges and benefits of shaker placement optimization for MIMO testing through a case study with the base of the Box Assembly with Removable Component (BARC). We first develop specifications for a random vibration test using measured field data. Next, we create an FEA model of the test assembly and calibrate this model with findings from an impact test. We use the FEA model and specifications to conduct an optimization that minimizes shaker force while matching the specification at chosen control DOFs. We then implement two laboratory tests: One using engineering judgment for determining shaker placement and one using the optimized shaker DOFs. Finally, we compare the control performance, required shaker force, and global system response between the two test configurations. Figure 1 outlines our approach for conducting this case study. The results indicate that despite using a calibrated FEA model on a relatively simple system, differences in system FRFs between model and hardware led to worse test performance from the optimized configuration.

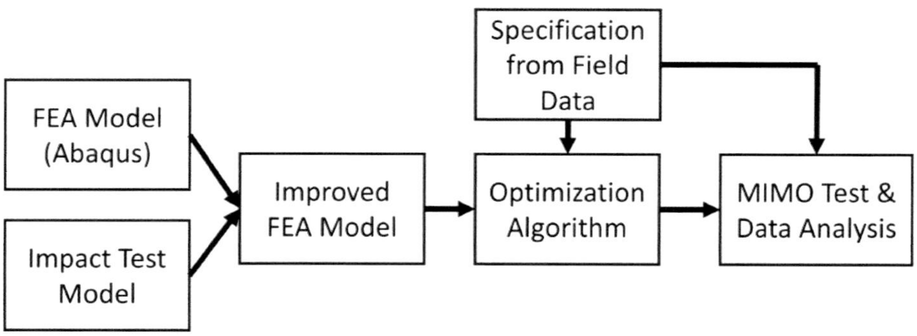

Fig. 1 Flowchart of steps associated with this project.

Environmental Specification

Developing a representative environmental specification is an important step for vibration testing. An overly conservative specification can lead to the accumulation of fatigue damage on potentially irreplaceable components. Conversely, an underestimate of the vibration environment can lead to critical modes not being excited to the level they would in normal operation. There are several methods for defining a vibration environment depending on the type of loading and the goal of the test. This study focuses on the random vibration associated with a transportation environment.

As noted in the introduction, we used the BARC base as a test article in this study. We instrumented the BARC with triaxial accelerometers, bolted the BARC base to an aluminum plate, and bolted the plate to a metal pallet in a truck bed. The truck was driven around paved and gravel roads for several minutes. We selected two accelerometer channels, 97Z+ and 58X+, to be used as control DOFs for the environmental test. Figure 2 shows the field test setup and sample time data from the two chosen accelerometer channels.

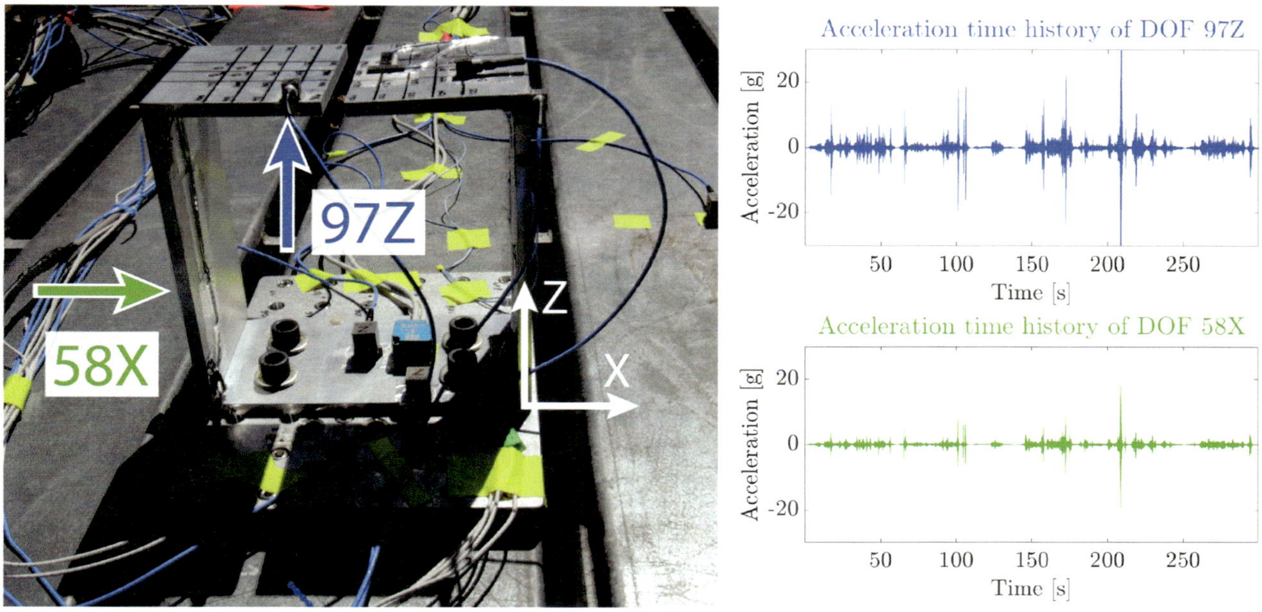

Fig. 2 For field data collection, the BARC base and plate were bolted to a pallet in a truck bed. DOFs 97Z and 58X were chosen as control DOFs, shown on the left, and sample time data from the field test are shown for each DOF on the right.

Random vibration tests are typically controlled to power spectral densities (PSDs) in the frequency domain, which describe the energy present in a signal at each frequency line. Given a time signal $x(t)$, let T denote the period of the signal and h_T denote a function that is equal to one within an arbitrary period and zero elsewhere. Computing $x_T(t) = x(t)h_T(t)$, we produce a signal that is equal to $x(t)$ for one period and zero elsewhere. Let $\hat{x}_T(\omega)$ denote its Fourier transform, with ω indicating a frequency variable. Then, the auto PSD $S_{xx}(\omega)$ is given by

$$S_{xx}(\omega) = \lim_{T\to\infty} \frac{1}{T}|\hat{x}_T(\omega)|^2. \quad (1)$$

To conduct our MIMO environmental test, we developed auto PSD specifications at each of our chosen control DOFs. Cross PSDs are also a consideration in MIMO tests, but we used an in-situ method to define the cross PSDs, described in the Environmental Testing section. Although we did not use the derived cross PSDs from the field data in this study, we did compute them. The Appendix provides more details on the computation and results of the derived cross PSDs.

It is evident from the time data in Figure 2 that the field data contained some minor shocks. Impulses that exceed the average amplitude of the random vibration data would negatively impact the auto-PSD estimation. Furthermore, we planned to use our test software in the Random Vibration configuration, where it is capable of generating only stationary Gaussian-distributed random vibration [6]. Therefore, the data used to generate the specifications needed to match those characteristics [7]. We excluded the shocks present in the data by dividing the time signal into 0.2-second blocks and accepting or rejecting each block using a root-mean-square (RMS) threshold. The value of this threshold was computed based on a P70/50 upper tolerance limit estimation of the RMS amplitudes for each signal block. The 70/50 refers to 50 percent confidence in a 70th percentile upper tolerance limit for the estimated RMS.

Notably, we fit the RMS amplitudes to a normal distribution in the log domain. Working in the log domain allowed for a better fit to a normal distribution, although other distributions were explored. Selection of optimal distributions for RMS rejection is outside the scope of this work, so we decided to primarily focus on normal distributions. These RMS values correspond to the total energy in the time domain signal, so the energy from the shocks would weight individual blocks over the threshold. For a discrete time domain signal $x(t)$ with a set of n values $x_1, x_2, \cdots, x_n$, the RMS is defined as

$$x_{RMS} = \sqrt{\frac{1}{n}(x_1^2 + x_2^2 + \cdots + x_n^2)}. \tag{2}$$

We chose time blocks of 0.2 seconds to keep as much of the testing data as possible, therefore producing a higher level of statistical confidence in our estimate. Figure 3 shows the time signal for DOF 97Z+ and DOF 58X+ with accepted and rejected blocks.

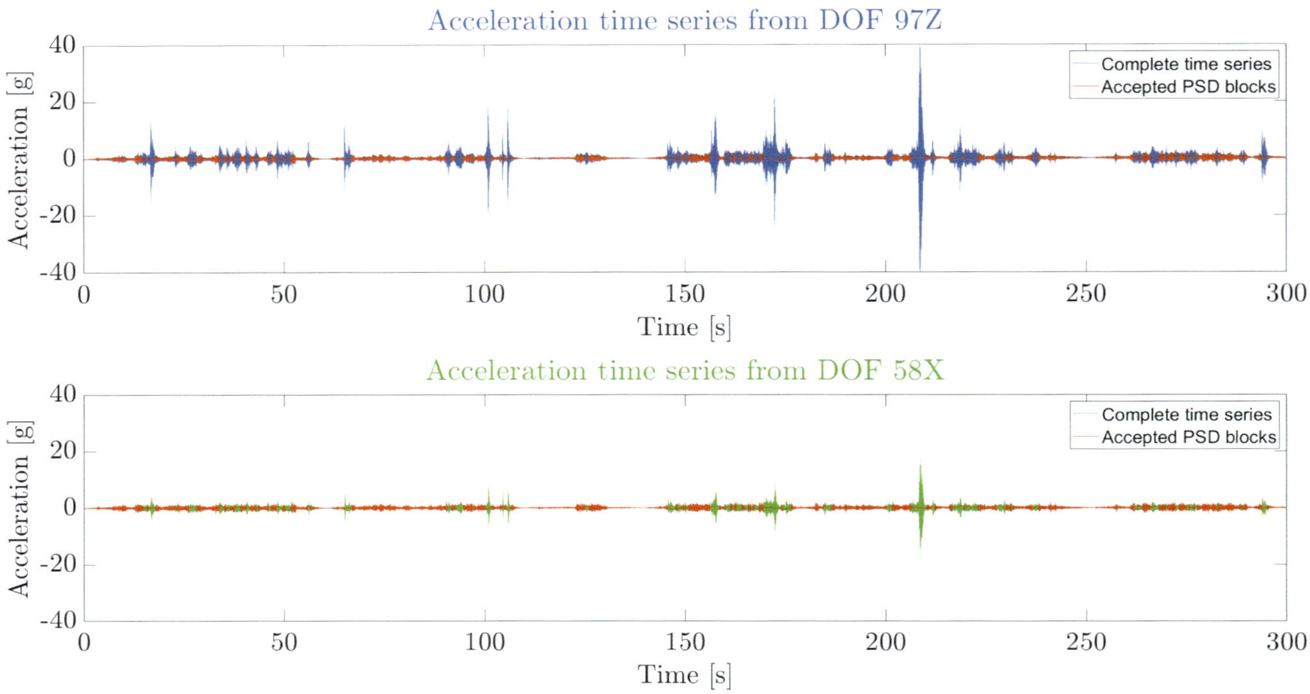

Fig. 3 Time signal for DOF 97Z and DOF 58X with accepted time series blocks.

For each accepted time block, we computed the auto PSDs for DOF 97Z+ and DOF 58X+. We used Welch's power spectral density estimate with the same parameters for each block so that the PSD of each block had values at the same frequency points. In the PSD computation, we used a Hanning window of length 160 points, 50% overlap between segments, and a 160-point discrete Fourier transform. This choice of parameters led to frequency lines with 10 Hz spacing. After examining the auto PSDs, we determined that a test range of 10-800 Hz would be sufficient for our environmental test. Under 10 Hz, the high-displacement vibration could prove a challenge for the small shakers used in this study. Above 800 Hz, the auto PSDs dropped below the noise floor of the accelerometers. Figure 4 shows an example PSD from DOF 97Z, plotted from 10-800 Hz.

We next developed a P95/50 maximum predicted environment (MPE) for each control DOF's auto PSD specification [8]. The 95/50 refers to 50 percent confidence in a 95th percentile upper tolerance limit for the estimated experimental auto PSD. This combination of confidence and tolerance is referred to as a Maximum Predicted Environment (MPE). We assumed a normal distribution for this work because the controller in our laboratory test uses normally-distributed excitation [6] [7]. First, we computed the sample mean $\bar{x}$ and sample standard deviation $\bar{s}$ across N samples. These values are then used to compute the K-factor k_1, an adjustment to the tolerance limit of our normal distribution:

$$a = 1 - \frac{\chi_\alpha^2}{2(N-1)} \tag{3}$$

$$b = \chi_p^2 - \frac{\chi_\alpha^2}{N} \tag{4}$$

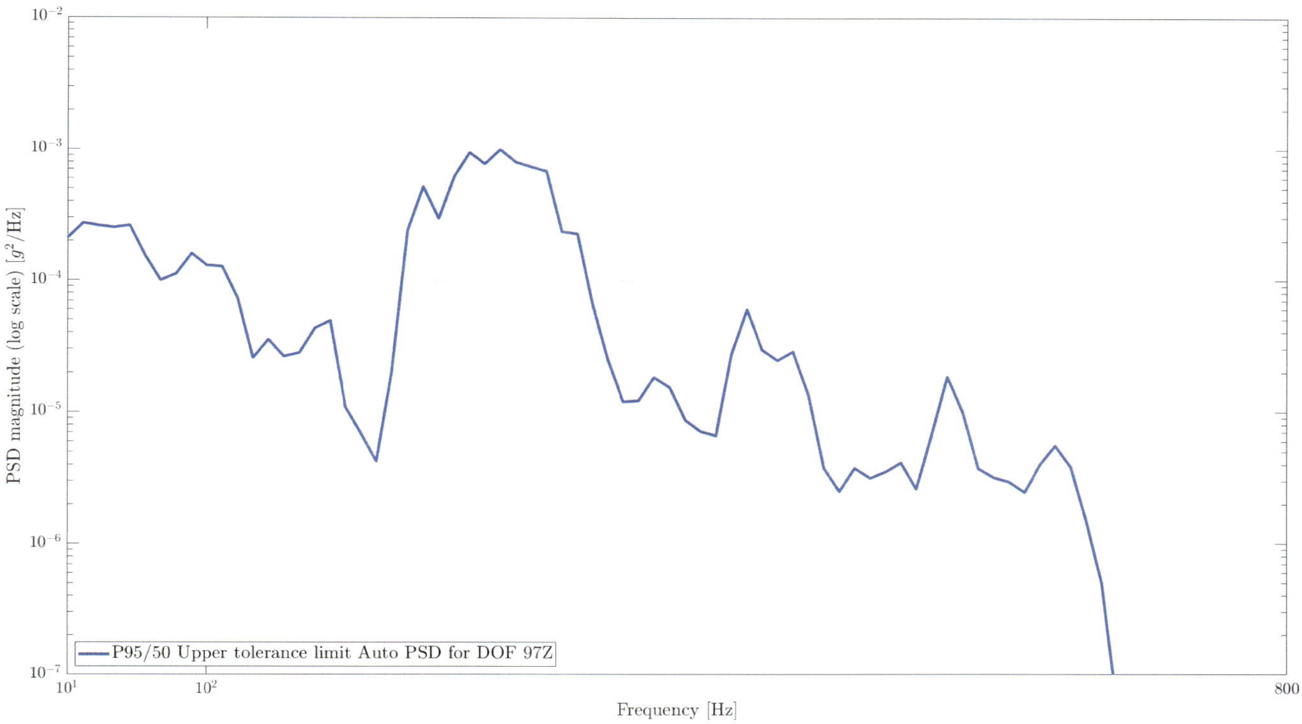

Fig. 4 Representative auto PSD of DOF 97Z from 10 to 800Hz. The drop-off after 700 Hz is due to application of a low-pass digital filter.

$$k_1 = \frac{\chi_p + \sqrt{\chi_p^2 - ab}}{a}, \tag{5}$$

where p is the desired tolerance limit and χ_α is the inverse cumulative distribution function (CDF) of the normal distribution associated with the desired confidence level α with a mean of $\overline{x}$ and a standard deviation $\overline{s}$. The upper tolerance limit Y_U is then given by

$$Y_U = \overline{x} + k_1\overline{s}. \tag{6}$$

Figure 5 shows an example histogram for one control DOF at one frequency line. Figure 6 shows the histogram for the same DOF across all frequency lines.

Finally, we smoothed the auto PSD upper tolerance limit from each channel across several frequency octaves to improve test implementation and to allow the environment to be more generalized [9]. At each octave center point, we took the mean of MPE P95/50 values within one half of an octave. This mean value was used to define the smoothed auto PSD. We interpolated on a log-log plot to find intermediary specification points between octave center point frequency values at a 1 Hz frequency spacing. The PSD vectors were used as the final specification by the MIMO vibration controller and our optimization algorithms.

The final auto PSDs are shown in Figure 7, which also displays the GRMS of the octave-smoothed signal in the title of each subplot. The GMRS, or acceleration RMS in units of earth's gravity g, is a metric describing the energy associated with a given signal. This metric is commonly used in vibration testing to denote relative amplitudes of different random vibration signals. In the frequency domain, the GRMS can be interpreted as the square root of the area under the curve. More formally, the GRMS can be computed from discrete frequency domain signal with a set of M values, given by $\{x_1, x_2, \cdots, x_M\}$ at corresponding frequency lines $\{1, 2, \cdots, M\}$ with frequency spacing f_s:

$$\text{GRMS} = \sqrt{\sum_{i=1}^{M} x_i f_s} \tag{7}$$

We note that by the GRMS metric, each DOF exhibited similar levels of energy in the field test, with DOF 58X slightly higher.

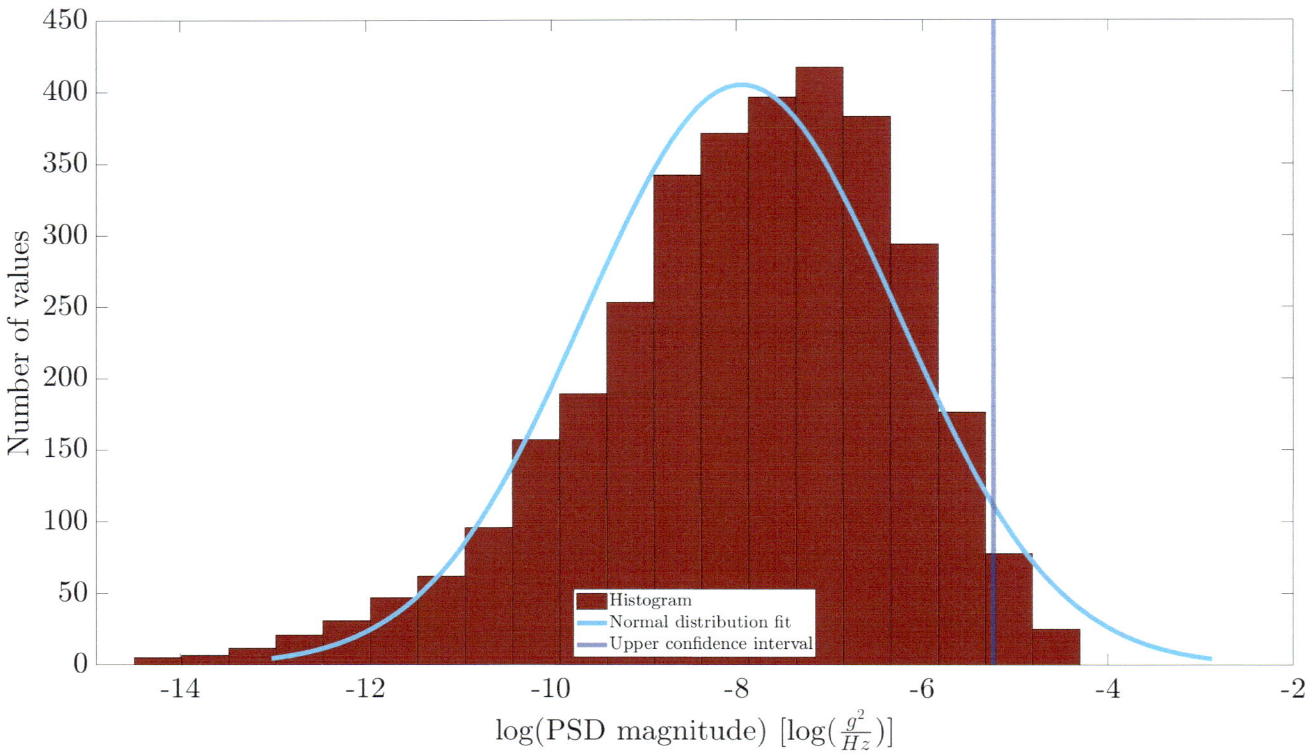

Fig. 5 PSD Magnitude histogram of DOF 97Z at 340Hz.

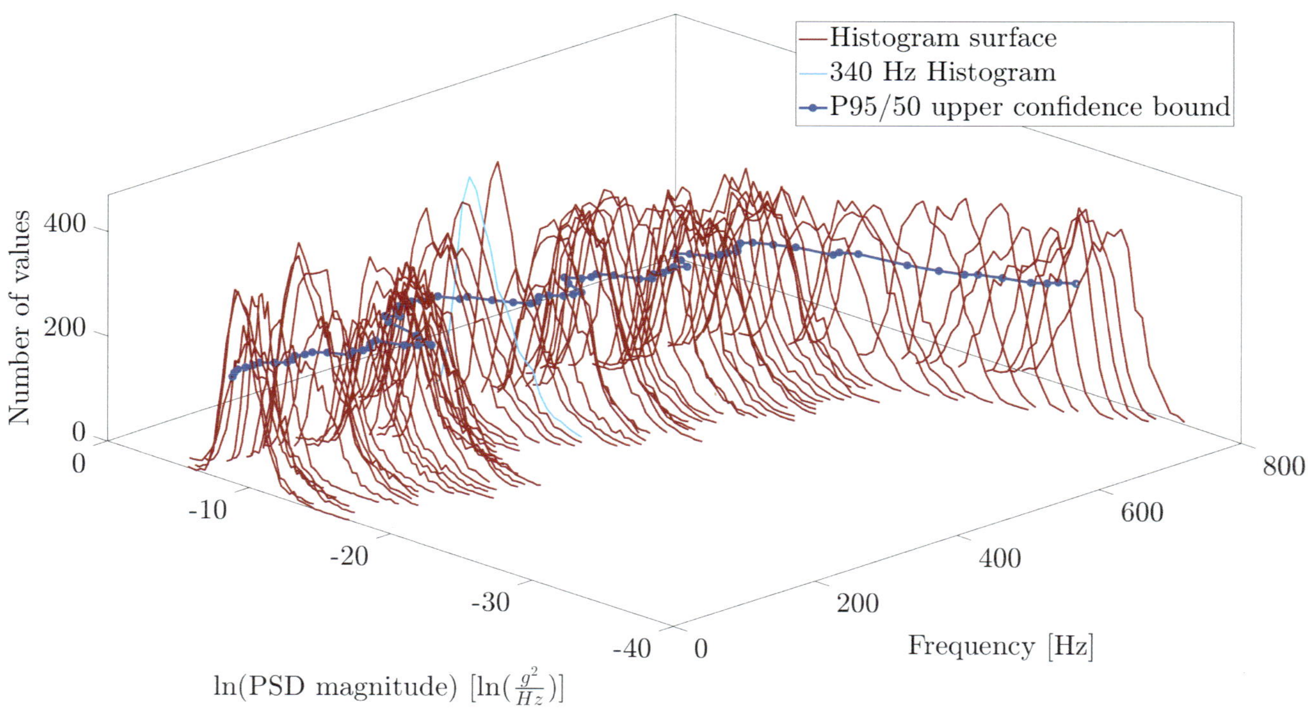

Fig. 6 PSD Magnitude histogram of DOF 97Z across all frequency lines.

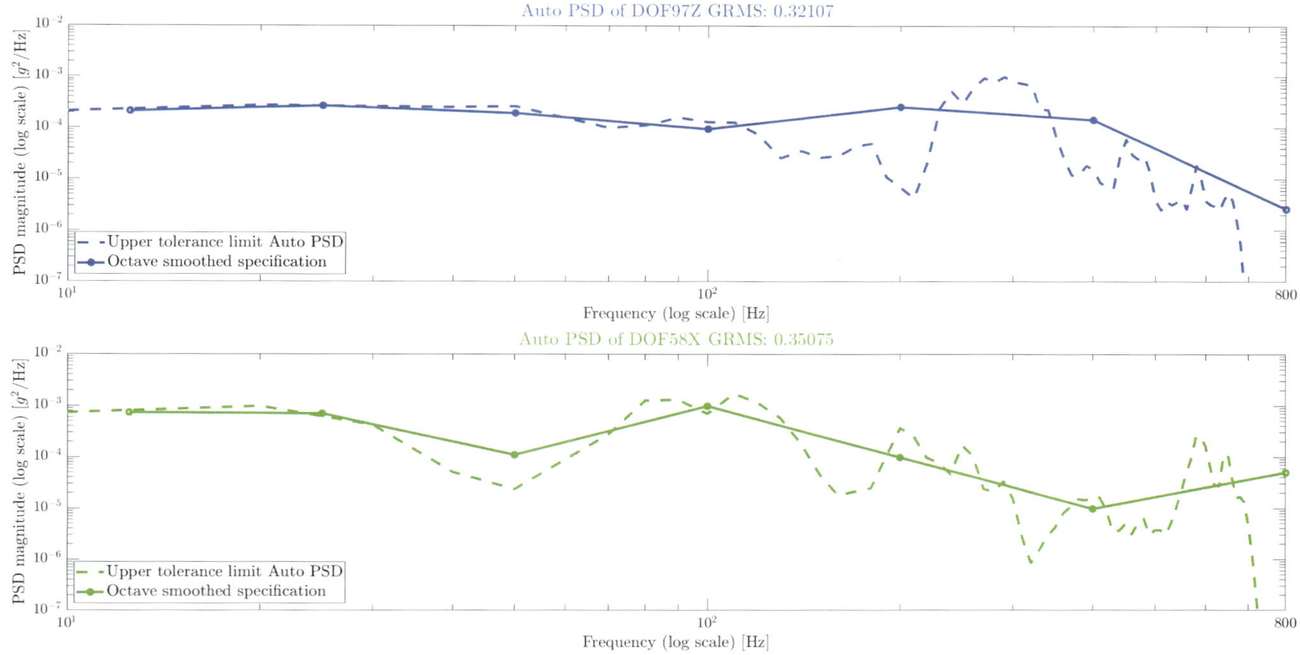

Fig. 7 Final MPE specification Auto PSDs for DOF 97Z and 58X.

Finite Element Analysis and Experimental Modal Analysis

With the specifications created for our environmental test, we next developed and refined a finite element analysis (FEA) model of the BARC base and plate. We created this FEA model to obtain a full set of accelerance frequency response functions (FRFs) for the shaker placement optimization algorithm. We chose a shell model rather than a solid model because previous work found that a shell model of the BARC was less computationally expensive while also producing less error between FEA and experimental modal analyses [10]. The material properties were nominal values for Al6061 (Aluminum), and we kept them consistent for both parts of the assembly.

We used existing engineering drawings and measurements with a ruler and caliper to create the geometry of the BARC base and plate. The shell model did not allow us to vary the diameter of the holes with depth, and since the plate holes included two different diameters, we took the average of them for our modeling. To model the connection between the BARC and plate, our first iteration applied a tie constraint between the circumferences of the bolt holes on the BARC and plate, as seen in Figure 8.

We then performed a mesh convergence study to determine the model's mesh size. We conducted a modal analysis in Abaqus while varying the global seed size from 1/2" to 1/128", reducing the size by a factor of two each time. Once we had computed the system's natural frequencies for each mode of each mesh size, we used Richardson Extrapolation to calculate the theoretical natural frequency of each mode [11]. To determine which mesh to use, we compared the error between the computed natural frequencies from each mesh size to the theoretical natural frequencies. We also considered the computation time for each simulation.

As we reduce the mesh size, we increase the fidelity of the model and expect the frequency values to begin to converge to the theoretical natural frequencies. However, consulting Table 1, we notice that for the lower-frequency modes, the frequency values begin to fluctuate instead of converging around a global seed size of 1/8" and 1/16". We believe this phenomenon is because the mesh becomes fine enough to reach the Abaqus program's noise levels. It is also important to note the simulation time. Computation time is roughly constant down to a global size of 1/16", after which it begins to increase exponentially.

The combination of computation time and fluctuation of natural frequencies prompted us to choose a mesh size of 1/32". The simulation time for this mesh size (81 s) was low enough to allow us to perform many iterations within the study time frame, and the simulated natural frequencies had an error of approximately one tenth of a percent or less when compared to the theoretical natural frequencies from the Richardson Extrapolation (Table 2).

Once we selected our mesh size, we proceeded to simplify the geometry of the assembly while maintaining the original dynamics of the system. We performed a modal analysis in Abaqus for several geometry simplification cases: (A) model the

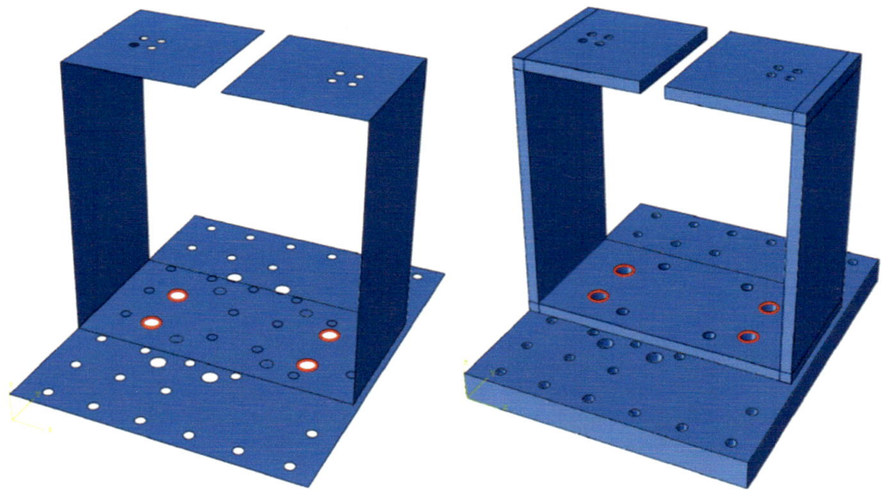

Fig. 8 Initial shell model of the BARC base and plate without the thickness visualized (left) and with the thickness visualized (right).

Table 1 Natural frequencies obtained from the mesh convergence study for each mesh seed size (1/128”-1/2”) and the simulation time.

	Natural Frequencies (Hz)						
Flexural Mode	**Global 1/128”**	**Global 1/64”**	**Global 1/32”**	**Global 1/16”**	**Global 1/8”**	**Global 1/4”**	**Global 1/2”**
1	101.15	101.16	101.17	101.14	101.19	101.17	101.33
2	153.45	153.45	153.45	153.42	153.47	153.45	153.66
3	254.54	254.63	254.80	254.82	255.66	256.03	257.07
4	367.11	367.19	367.34	367.44	368.25	369.03	371.26
5	514.58	514.61	514.69	514.70	515.13	515.39	517.13
6	525.93	526.01	526.18	526.29	527.41	528.83	532.34
7	574.80	574.82	574.89	574.92	575.34	575.71	577.86
8	583.85	584.01	584.34	584.29	586.36	586.92	589.41
Sim Time (sec)	2308	236	81	42	35	33	33

Table 2 Error between Richardson Extrapolation and the 1/32” mesh seed size for each flexural mode.

Flexural Mode	**Richardson Extrapolation Theoretical Natural Frequencies (Hz)**	**Error between 1/32” and Richardson Extrapolation (%)**
1	101.15	0.02
2	153.45	0.00
3	254.53	0.11
4	367.09	0.07
5	514.57	0.02
6	525.90	0.05
7	574.79	0.02
8	583.78	0.10

geometry of all holes in the assembly, (B) Remove all small holes on the plate, (C) Remove all small holes in both parts of the assembly, and (D) Remove all holes except the four main bolt holes connecting the BARC base to base plate. The respective geometries that we considered can be seen in Figure 9. Looking at the results of the FEA modal analysis in Table 3, we

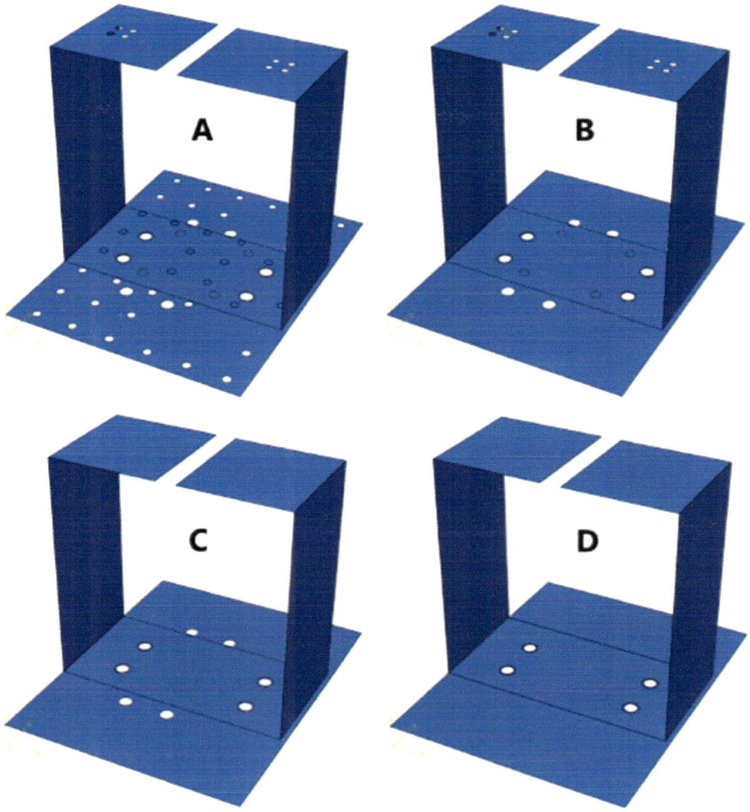

Fig. 9 Different BARC approximations from most detailed (A) to most simplified (D).

Table 3 Natural frequencies for each geometry simplification for the 1/32" mesh size.

Flexural Mode	Natural Frequencies (Hz) A	B	C	D	Error between A and D (%)
1	101.13	101.17	100.86	100.86	0.27
2	153.92	153.41	153.08	153.02	0.58
3	254.41	254.85	254.38	254.44	0.01
4	368.68	367.24	367.06	366.84	0.50
5	514.86	514.68	515.24	515.17	0.06
6	526.19	526.20	526.38	526.41	0.04
7	575.90	574.80	575.74	575.63	0.05
8	586.40	584.20	584.51	584.18	0.38

notice that the percent difference of the natural frequencies between our most simplified model (D) and the original model (A) are approximately half a percent or less. Given the low difference, we chose option D and simplified the geometry.

After developing the preliminary model, we performed an experimental modal analysis of the assembly using a roving hammer impact test. We suspended the BARC and plate from an aluminum frame using rubber bands and zip ties to simulate the free-free boundary condition which was used in the FEA. We then attached five uniaxial accelerometers to our assembly, with two in the system X and Z axes and one in the Y axis. The accelerometers were exclusively on the BARC component of the assembly because we expected most of the motion to be there based on FEA simulation results and engineering judgment. We then outlined a grid on the assembly consisting of 178 nodes. Ideally, each node would be impacted in all three axes; however, due to the assembly's geometry, some DOFs were impossible to excite. Furthermore, we expected some DOFs to cause the system to act identically due to the symmetry of our assembly. With this in mind, we reduced the number of nodes to 171 and the DOFs to 215. The experimental setup is shown in Figure 10.

Fig. 10 Two photos of the test setup for experimental modal analysis.

We conducted the testing using Siemens SCADAS Data Acquisition hardware and Simcenter Testlab software, impacting each of the 215 DOFs five times with an instrumented hammer. We processed the impact test data using IMAT, a Matlab toolbox for modal analysis by ATA Engineering. We fit the FRFs using an orthogonal polynomial fitting technique and used these FRFs to obtain natural frequencies and mode shapes of the experimental system.

During the initial comparison between experimental and simulation modal test results, we saw high errors in natural frequency values and observed a mode missing from the experimental test that was present in the FEA results. In reflecting on the large error, we realized that we had made a strong assumption regarding how the plate and the BARC are connected in Abaqus. We also understood that the nonlinear behavior of the contact between the BARC and the base plate would be difficult to model with a linear modal analysis, thus we began to explore various constraint schemes (contact areas) in an effort to best match the experimental modes observed in the impact test. After approximately 40 iterations of tie constraints, we obtained a reasonable result with an average error of 3.13% across the seven modes found in the experimental analysis. Furthermore, the FEA mode shapes qualitatively matched the experimental mode shapes, and the extra mode that was present in the original FEA mode also disappeared with this new model. Note that the number of modes in the frequency range of interest reduced to seven.

Table 4 Natural frequencies for the updated FEA model, the hammer test, and the error between them.

	Natural Frequencies (Hz)		
Flexural Mode	**Hammer Test**	**Updated FEM**	**Error (%)**
1	121.38	117.51	3.19
2	177.13	170.29	3.86
3	315.99	330.64	4.64
4	405.94	402.57	0.83
5	571.44	563.16	1.45
6	643.39	627.27	2.51
7	700.63	662.72	5.41
		Avg Error	**3.13**

The final model's tie constraint is shown in Figure 11. In this final model, we removed all bolt holes, as it simplified the FEA model and produced a lower error on the natural frequencies when compared to experimental results. We attempted other modifications, such as altering BARC and plate thickness and slightly varying the density and modulus of elasticity, but they proved to be ineffective at correcting the frequency errors and the mode shapes.

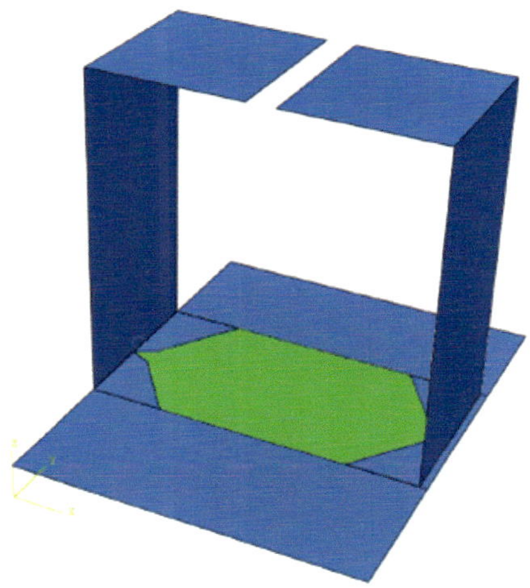

Fig. 11 The final FEA model with the new tie constraint (green area).

Once we were satisfied with the relatively low error as well as the qualitative mode shapes, we performed a Modal Assurance Criterion (MAC) to quantitatively assess mode shape differences between FEA and experiment. The MAC is shown in Figure 12. Modes 1-6 quantitatively match well, but mode 7 has low similarity between FEA and experiment. Figure 12 shows a visual comparison of mode 7 shapes between experimental and FEA analysis. We believe the low correlation could have resulted from poor excitation during the experimental hammer test. However, because mode seven is at 700 Hz, around when the energy in the specifications drops (Figure 4), we decided to proceed with the optimization using this FEA model.

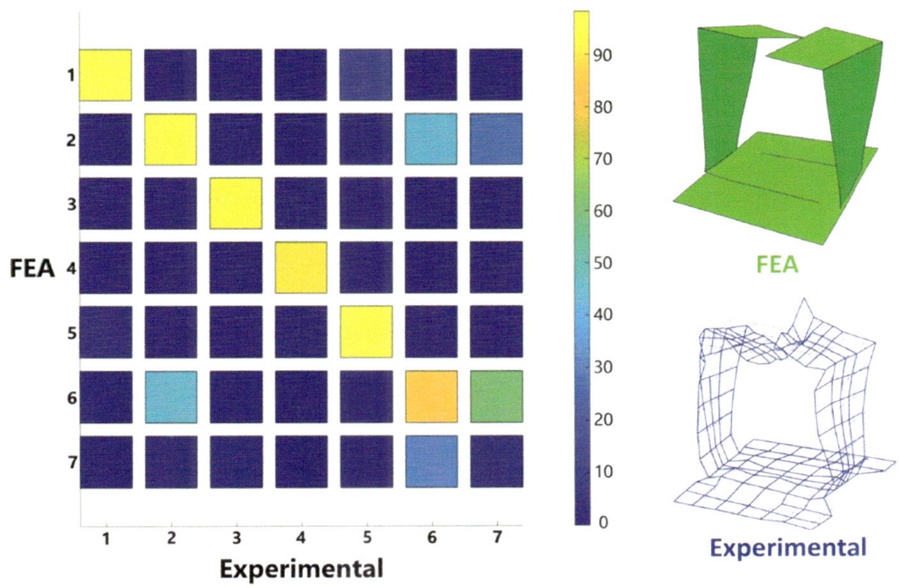

Fig. 12 Modal assurance criterion between the FEA and experimental mode shapes (left) and a visual comparison of mode 7 displacement between FEA and experimental analysis (right).

Shaker Placement Optimization

Distributed shaker vibration tests often pose challenges because the modal shakers used for excitation have a relatively low force input compared to the large shakers used for single-axis testing. Although the environment developed in the Environmental Specification section is a low-level transportation environment, we aimed to develop a testing pipeline that could be generalized for other environments, such as flight. We therefore chose to optimize shaker locations and directions that minimize force to achieve a specified environment. This paper describes an optimization for two shakers because the hardware setup described in the Environmental Testing section only allowed for two outputs. Our optimization aimed to minimize the input force spectrum required by the two shakers to achieve the specification at the two chosen control DOFs.

We imported the geometry, natural frequencies, and mode shapes from the updated FEA model into SDynPy, a structural dynamics Python toolbox developed at Sandia National Laboratories[12]. Each node in the finite element model has three DOFs, which produced too large of a search space for the optimization algorithm. We reduced the search space by selecting a subset of nodes in a 0.25-inch grid, then removing DOFs that were not feasible locations to place a shaker. These candidate DOFs can be seen in Figure 13.

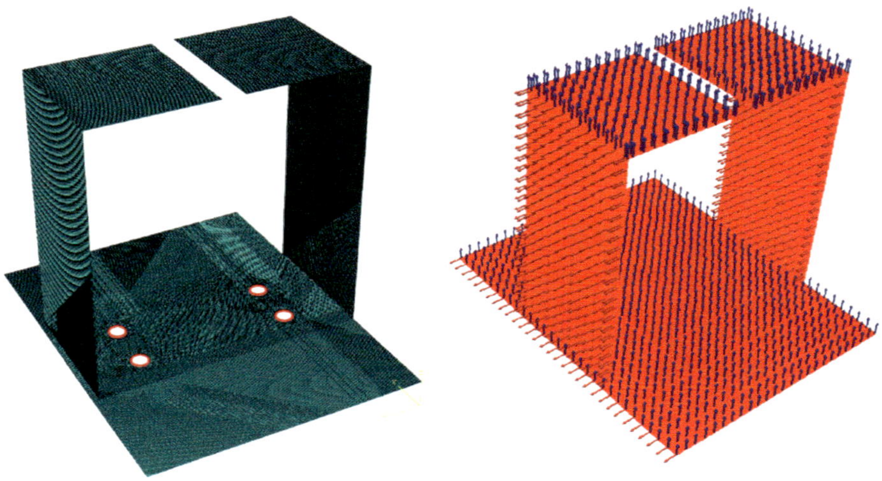

Fig. 13 FEA Model (left) and final set of candidate DOFs with 0.25-inch spacing (right).

Using the system geometry, natural frequencies, and mode shapes in SDynPy, we synthesized FRFs for the BARC system. We imported the MPE auto PSDs developed in the Environmental Specification section, defined as $S_i = [S_{1,i}\ S_{2,i}]^T$ at control DOFs 1 and 2 for each frequency line i ($i \in \{1, 2, \cdots, M\}$ for M frequency lines). At each iteration of the optimization, the algorithm selects two candidate DOFs, given by j and k, as shaker locations. It then computes the synthesized FRF matrix H_i, a 2×2 matrix between the input DOFs (j, k) and the control DOFs $(1, 2)$ for each frequency line i. We use H_i and S_i to compute the required input force spectra $F_i = [F_{j,i}\ F_{k,i}]^T$ to achieve the specification. Our optimization formulation is then given by

$$\begin{aligned} \min_{j,k} \quad & \sqrt{\sum_{i=1}^{M} F_{j,i} f_s} + \sqrt{\sum_{i=1}^{M} F_{k,i} f_s} \\ \text{s.t.} \quad & S_i = H_i F_i \\ & i \in \{1, 2, \cdots, M\} \\ & j, k \in \{1, 2, \cdots, N\} \\ & j \neq k, \end{aligned} \tag{8}$$

where N is the number of candidate input DOFs and f_s represents the frequency spacing. Note that the objective function is the sum of force RMS values from each shaker, using the definition of RMS provided in Equation 7. As shown in Equation 8, we added constraints that require the set of candidate DOFs to be a valid set and prohibit the two shaker DOFs from being the same at each iteration.

We implemented this optimization formulation using two solvers from the Python SciPy package: *Brute Force* and *Differential Evolution*. The Brute Force solver searches through every possible combination of candidate DOFs. This strategy finds the globally optimal solution, but it is computationally inefficient. Differential Evolution works with a population of candidate input DOFs. The members of this population are moved around the search space. If a new DOF produces a lower objective function value, then it is accepted back into the population; if it increases the objective function value, then it is discarded. This process is repeated until a satisfactory solution is found. Differential Evolution does not guarantee a globally optimal solution.

Because the optimization used FRFs synthesized from our FEA model, the node numbers in the optimization solution are reported in the FEA model numbering scheme. The Brute Force algorithm produced the candidate set 7X+ and 1929Z+, whereas Differential Evolution found 1937Z+ and 609X+. These DOFs are shown in the middle and right panels of Figure 14, where it can be seen that these solution sets are quite similar. The time to compute these sets, however, differed greatly, with Brute Force taking 10,800 seconds and Differential Evolution taking 73.1 seconds. When mapped onto the coarser experimental grid used for experimental modal analysis, the solution was the same for either solver. Given the similarity in the sets found, we recommend proceeding with Differential Evolution in future work.

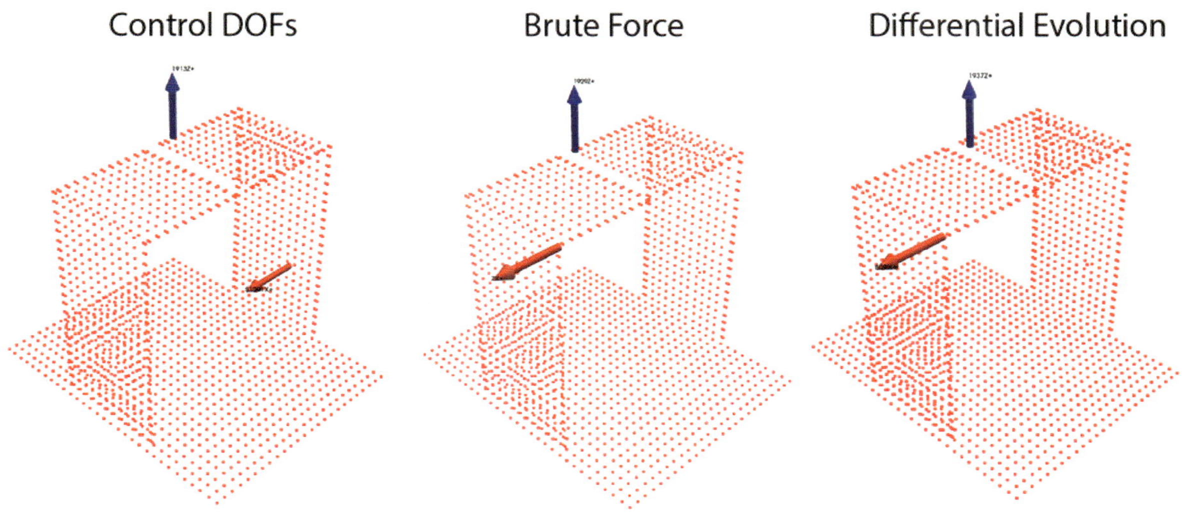

Fig. 14 Control DOFs (left), optimal candidate DOFs from Brute Force (middle), and optimal candidate DOFs from Differential Evolution (right).

Environmental Testing

For environmental testing, we slightly altered the test setup from the impact test configuration. The original rubber bands used for the impact test could not be used due to stretching, so they were replaced. We selected replacement rubber bands based on approximate similarity to the original rubber bands because we aimed to replicate the system that was characterized in the impact test. Similarly, two shakers (The Modal Shop K2007E01) were suspended from the test frame on bungees to decouple them from the flexible modes of the test stand. This decoupling is particularly important in the context of shaker placement optimization because our optimization algorithm did not account for the interaction between shaker and frame as a function of location.

We implemented random vibration MIMO tests in two configurations for this study. The first configuration used engineering judgment to collocate the two shakers with the two control DOFs. Intuitively, this configuration reflected the optimization priorities of matching the auto PSD at the two control DOFs while minimizing shaker force. The second configuration used the shaker DOFs found by the optimization algorithm. Figure 15 shows the configuration for each test. A gain of 18dB was prescribed for the shakers during both test configurations.

We used the Rattlesnake Vibration Controller software, an open-source Python software developed by Sandia National Laboratories, coupled with National Instruments data acquisition hardware to control multiple shakers during testing [13]. For all tests, we used a built-in controller called *buzz control* with a 50% overlap Tukey window, a constant overlap and add (COLA) time series generator, and a 20-frame (average of 20 seconds) dynamic spectral density matrix estimation.

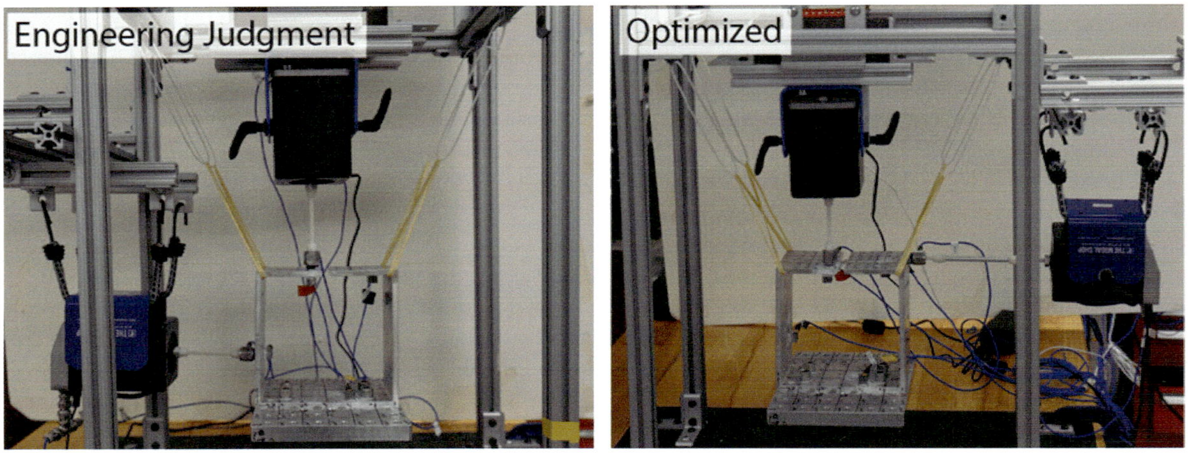

Fig. 15 Hardware test configurations for the engineering judgment setup (left) and optimized setup (right).

Sampling was set at 1800 Hz with a frequency line spacing of 1 Hz to align the frequency lines of the dynamic estimation with that of the specification. The Rattlesnake controller performs a system identification step for each test to re-estimate transfer functions between shaker voltage and control DOF acceleration. The buzz control algorithm then uses the in-situ measurements from the system identification step to populate the cross powers in the specification. Test data were collected for 2 minutes for each test.

Analysis

Before comparing the two tested configurations (engineering judgment and optimized), it is useful to describe relevant error metrics and their relative importance. In our application of random vibration testing, the "targets" are control auto PSDs at specified DOF. Our target auto PSDs and measured test auto PSDs are vectors with one value per frequency line, but it is often useful to reduce test performance metrics to a single value for easier comparison [2]. We therefore consider vector error metrics described below as well as comparing RMS values between testing configurations.

In the Environmental Specification section, we described computing MPE auto PSDs for channels 97Z and 58X. We followed this same process to compute MPE auto PSDs for all other measured acceleration channels in the field data, for 12 auto PSDs total. These PSDs will be referred to as *field auto PSDs*. To compute a global error metric for all measured channels, we define the field auto PSD values as $s_{p,i}$, where p indicates the channel number ($p \in \{1, 2, \cdots, N\}$) and i indicates the frequency line ($i \in \{1, 2, \cdots, M\}$). Likewise, we define experimental auto PSD values as $\overline{s_{p,i}}$. For this study, we used 791 frequency lines and 12 acceleration channels, but the derivations will be presented for general N and M here. We concatenate these auto PSD values into $N \times M$ matrices $S_{N,M}$ and $\overline{S_{N,M}}$, defined as

$$S_{N,M} = \begin{bmatrix} s_{1,1} & \cdots & s_{1,M} \\ \vdots & \ddots & \vdots \\ s_{N,1} & \cdots & s_{N,M} \end{bmatrix}_{N \times M} \tag{9}$$

$$\overline{S_{N,M}} = \begin{bmatrix} \overline{s_{1,1}} & \cdots & \overline{s_{1,M}} \\ \vdots & \ddots & \vdots \\ \overline{s_{N,1}} & \cdots & \overline{s_{N,M}} \end{bmatrix}_{N \times M} . \tag{10}$$

We then compute the relative error matrix $R_{N,M}$ using these two matrices. We chose relative rather than absolute error to avoid over-weighting the highest amplitude channel, as errors tend to be proportional to excitation amplitude. $R_{N,M}$ is given by

$$R_{N,M} = \begin{bmatrix} \frac{\overline{s_{1,1}}}{s_{1,1}} & \cdots & \frac{\overline{s_{1,M}}}{s_{1,M}} \\ \vdots & \ddots & \vdots \\ \frac{\overline{s_{N,1}}}{s_{N,1}} & \cdots & \frac{\overline{s_{N,M}}}{s_{N,M}} \end{bmatrix}_{N \times M} . \tag{11}$$

The error values in $R_{N,M}$ still have widely differing magnitudes at any given frequency line based on the magnitude of the specification, so we next normalize the error metric to avoid over-weighting the lowest amplitude channel [2]. We define a normalizing vector η of length M, where each entry is the L2 norm of the specification magnitude at that frequency line:

$$\eta = \left[\|S_{1,1\to N,1}\| \quad \cdots \quad \|S_{1,M\to N,M}\|\right]_{1\times M}, \tag{12}$$

where, for each frequency line i,

$$\|S_{1,i\to N,i}\| = \sqrt{s_{1,i}^2 + \cdots + s_{N,i}^2}. \tag{13}$$

We then compute a weighting factor $W_{N,M}$ to be applied to the relative error:

$$W_{N,M} = \begin{bmatrix} \frac{s_{1,1}^2}{\eta^2} & \cdots & \frac{s_{1,M}^2}{\eta^2} \\ \vdots & \ddots & \vdots \\ \frac{s_{N,1}^2}{\eta^2} & \cdots & \frac{s_{N,M}^2}{\eta^2} \end{bmatrix}_{N\times M}. \tag{14}$$

Next, we define the the weighted relative error matrix $E_{N,M}$ as element-wise multiplication of $R_{N,M}$ and $W_{N,M}$:

$$E_{N,M} = \begin{bmatrix} R_{1,1}W_{1,1} & \cdots & R_{1,M}W_{1,M} \\ \vdots & \ddots & \vdots \\ R_{N,1}W_{N,1} & \cdots & R_{N,M}W_{N,M} \end{bmatrix}_{N\times M}. \tag{15}$$

Lastly, we sum $E_{N,M}$ across the N channels to obtain the global error vector G:

$$G = \left[\sum_{p=1}^{N} E_{p,1} \quad \cdots \quad \sum_{p=1}^{N} E_{p,M}\right]_{1\times M}. \tag{16}$$

With error metrics defined, we next present test results. We first compared the force auto PSD GRMS values between the two test configurations (engineering judgment and optimized). Figure 16 plots the force auto PSDs and reports the lbRMS (RMS of the signal in lb) value for each shaker. The optimized configuration resulted in a lower force lbRMS

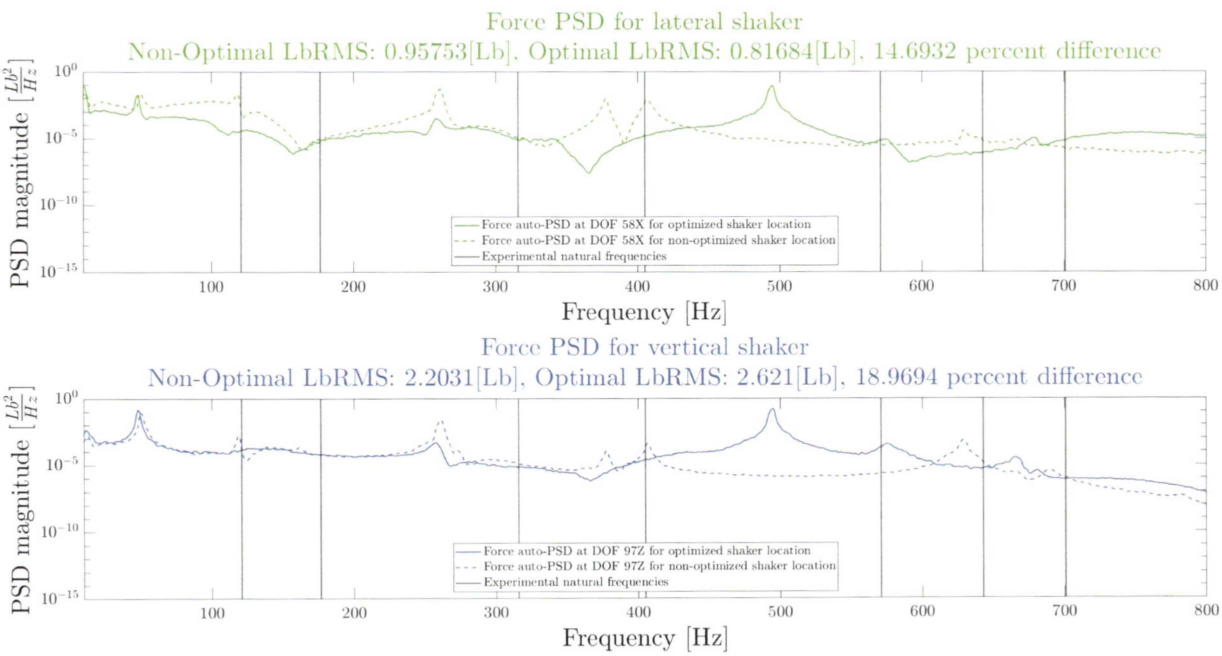

Fig. 16 Comparison of required force auto PSDs between optimized (solid) and non-optimized (dashed) shaker locations. The natural frequencies from experimental modal analysis are indicated by black vertical lines.

for the horizontal shaker, but a higher lbRMS for the vertical shaker. Averaging lbRMS results between both shakers, the engineering judgment test required less force than the optimized test.

After analyzing the measured force between configurations, we assessed measured acceleration in each test configuration. Figure 17 shows the auto PSDs at the two control locations compared to the specification for the engineering judgment shaker test. We observe the largest deviation from the specification around 350-400 Hz. The GMRS percent error between specification and test was 0.957% and 6.02% for the horizontal and vertical DOFs, respectively. Figure 18 shows the global

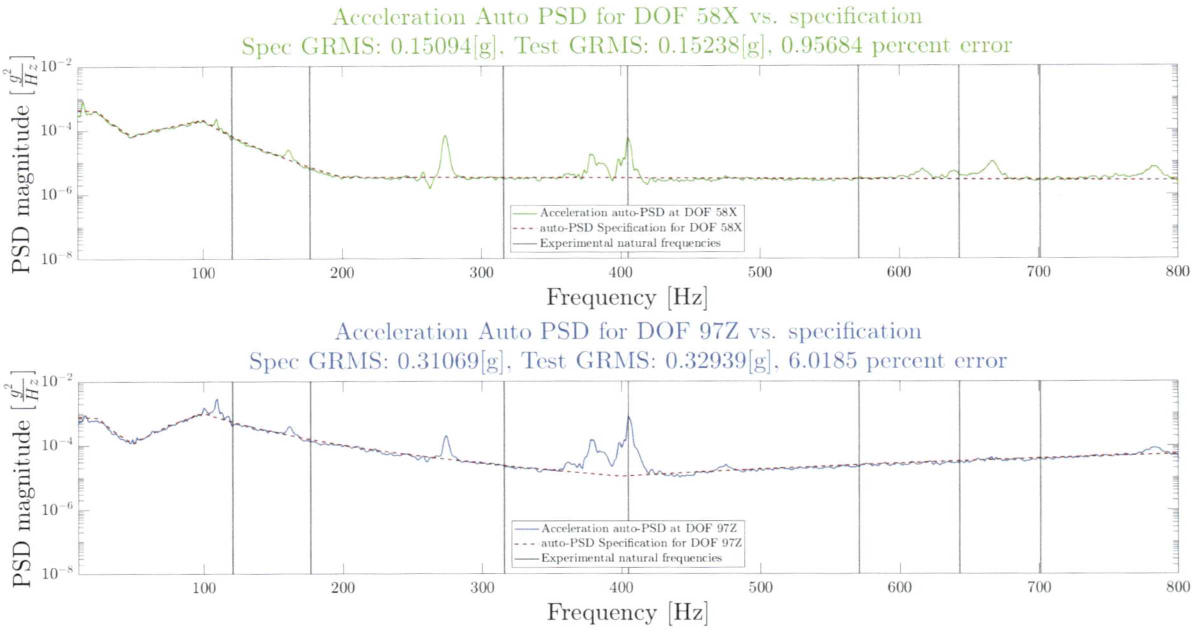

Fig. 17 Specification auto PSDs (red, dashed) and experimental auto PSDs (solid) for the engineering judgment test configuration. The natural frequencies from experimental modal analysis are indicated by black vertical lines.

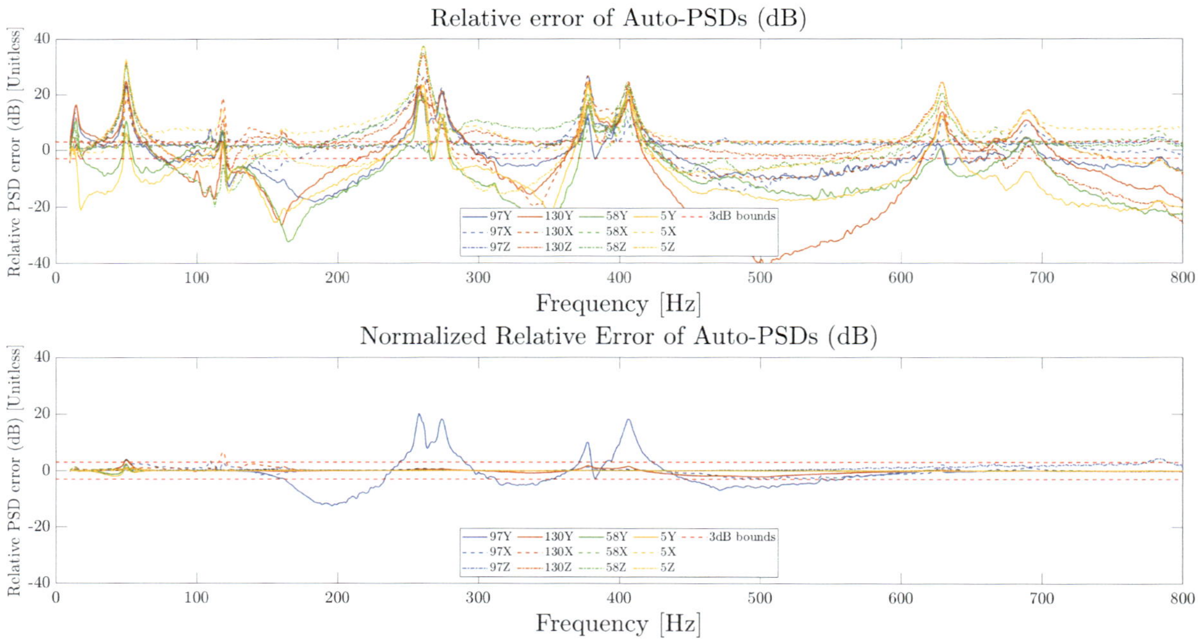

Fig. 18 Global relative error for all measured DOF for the engineering judgment test configuration.

relative error, both un-normalized and normalized, for all measured acceleration channels in the engineering judgment test. One interesting observation is that DOF 97Y has the highest normalized global error. This DOF could be an ideal control channel in future multi-input tests.

Next, we conducted the same analysis for the optimized test configuration. Figure 19 compares the auto PSDs between specification and test at the two control locations for the optimized shaker test configuration. Visually comparing these results to Figure 17, we observe a smaller deviation from the specification in the 350-400 Hz range, but a larger deviation from 700-800 Hz. The GMRS percent error between specification and test was 9.15% and 4.29% for the horizontal and vertical DOFs, respectively. By this GRMS metric, the optimized shaker location performed worse for the horizontal DOF and slightly better for the vertical DOF. Averaging across control channels, the engineering judgment test produced better control performance. Figure 20 shows the un-normalized and normalized global relative error for all measured acceleration channels in the optimized test. Just as with the engineering judgment test, DOF 97Y has the highest normalized global error.

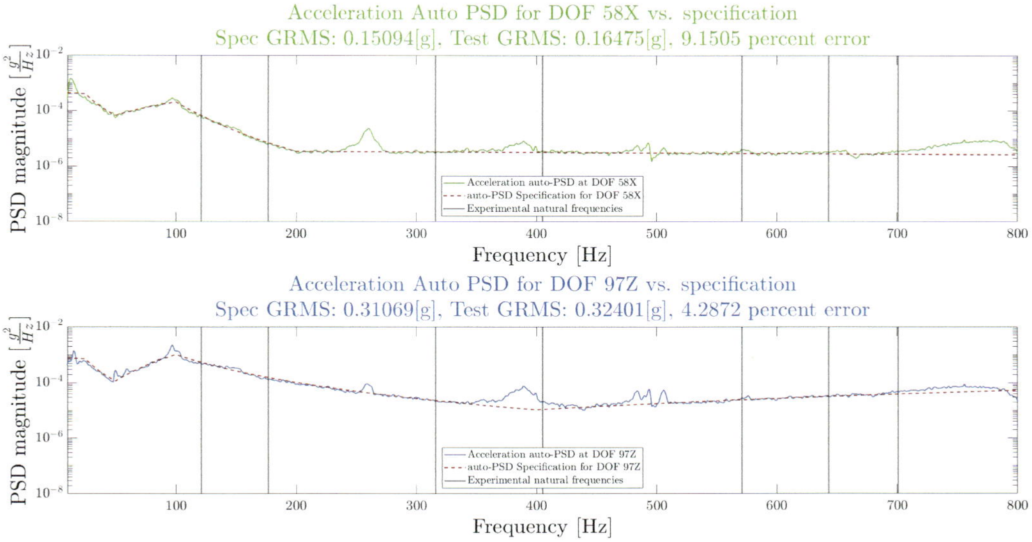

Fig. 19 Specification auto PSDs (red, dashed) and experimental auto PSDs (solid) for the optimized test configuration. The natural frequencies from experimental modal analysis are indicated by black vertical lines.

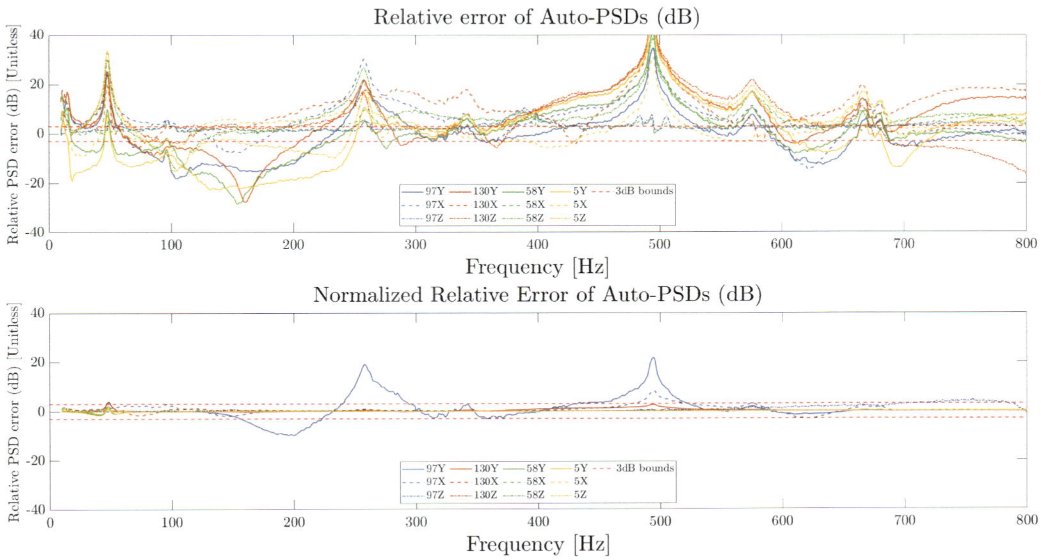

Fig. 20 Global relative error for all measured DOF for optimized shaker locations

Discussion

The optimization algorithm implemented in this study aimed to minimize shaker force while matching the specifications at the two control DOFs. However, our analysis revealed that the optimized shaker locations required more force than the engineering judgment configuration, as calculated by the sum of the lbRMS across the two shaker input force auto PSDs. This result was particularly surprising because the optimized shaker DOFs were located on the most flexible part of the structure. Additionally, the optimized shaker locations performed worse with respect to average error from test specification auto PSDs.

Potential sources of error for the force auto PSDs include a lack of modeling of the impedance associated with the shakers, lack of modeling of suspension stiffness, and shakers not being perfectly aligned during testing. Additionally, failing to model system damping in the FEA model may have contributed to increased error. Figure 21 shows sample FRFs between 58X and 97Z from the FEA model, experimental modal analysis, and environmental test. Visually, the addition of the shakers causes a large change in the FRF, contributing to the discrepancy between FEA optimization and hardware.

Notably, the first two modes were similar in frequency and magnitude across all three FRFs in Figure 21, but it is unclear whether this trend holds true across all the FRFs in the optimization search space. We were unable to run a comparison of all FRFs because we only tested two shaker test configurations. However, it is reasonable to assume that the addition of shakers in any test configuration would create differences between hardware and FEA FRFs. In Figure 21, the impact test and FEA FRFs are more similar, which is to be expected because we updated the FEA model to match impact test results. It is unclear how best to estimate the effect of shaker impedance during the optimization, but the discrepancy between the impact test FRFs and shaker test FRFs indicate that this consideration is crucial for effective shaker placement optimization.

The decrease in control performance with the optimized shaker locations (Figure 17 versus Figure 19) is similarly notable. The shakers used in this study had three gain settings. Depending on the transfer function between input voltage and output acceleration, any gain setting could simultaneously be too high for a given frequency line and too low for other frequency lines. For example, if an auto PSD specification at a given frequency line is 10^{-4} g^2/Hz while the transfer function magnitude at that frequency line is 10^{-6} g/V (as it might be at an anti-resonance), the gain would need to be set around 20 dB to keep the control signal between the desired bounds of $\pm 1V$. However, if that same auto PSD specification continued to a frequency line where the transfer function magnitude was 10^{-4} g/V (such as at a mode), a gain of 20dB would be excessive. The optimized shaker configuration produced different transfer functions between shaker and control DOFs, and our choice gain settings may have contributed to the over-test from 700-800 Hz, resulting in higher GRMS error.

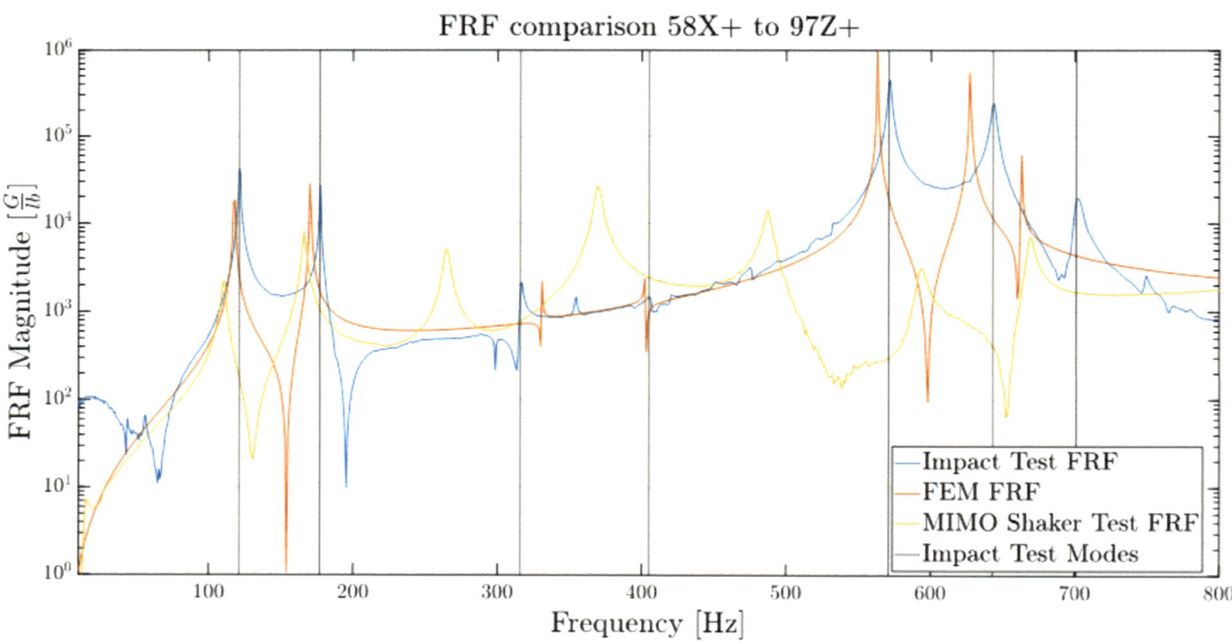

Fig. 21 FRFs computed between DOFs 58X and 97Z from the impact test, finite element analysis, and MIMO shaker test. The black vertical lines show the frequency values of the modes from the impact test.

Additionally, rigid body motion likely deflected shakers relative to the BARC and caused some bending in the stingers. This behavior could have created non-measured shear and moments applied to the structure, thus impacting the control performance. The rigid body motion could be addressed in future work through improved control algorithms for low frequencies, better experimental practices, or changing the testing range of interest.

Conclusions

The results from this study indicate that our optimization formulation, which minimized GRMS shaker force while matching the auto PSD specification at two control DOFs, was not effective when applied to the FEA model of the BARC base and plate. We found several reasons that explain the poor performance.

First, we observed differences in peak location and amplitude in the FRFs between FEA model, experimental modal analysis, and in-situ conditions during the environmental test (Figure 21). We certainly expected differences between model and hardware, but the discrepancy observed in this study was large enough to make an optimization with FEA FRFs effectively useless in the hardware test. We intentionally ignored the shaker impedance in the FEA model due to time constraints, but its omission led to an inaccurate system characterization.

Second, our transportation environment had a lower frequency bound of 10 Hz. The large displacements caused by low-frequency excitation most likely led to shear/moment transfer through the shaker stingers. The applied moments led to control error, which we believe is related to the increase in force. Across our tests, increased control error was correlated with increased lbRMS requirements.

Future work could investigate higher fidelity FEA modeling, such as the inclusion of shakers in the FEA model. The challenge with including shaker models in a shaker placement optimization is that the shaker locations and directions are not known a priori, so the system FRFs would have to be updated on each optimization iteration. However, this added computation may be worthwhile given the differences in FRFs between FEA and hardware observed in this study.

Another suggestion for future work is to explore changes in the search space of the optimization algorithm, such as incorporating non-normal excitation directions. Furthermore, the sensitivity of optimization results to spatial discretization in the search space could be explored. As the efficiency of search algorithms depends on the size of the search space, investigating what degree of model fidelity meaningfully contributes to results could aid in the application of shaker placement optimization to more complex assemblies than the BARC.

Additionally, new optimization problems could be explored. In this study, we notably did not consider the cross power spectra in the optimization problem. Our cost and constraints focused on matching the auto PSDs at control DOFs while minimizing force. Future work could consider the achievability of cross power spectra, include more control DOFs than shakers, or consider the response at non-control DOFs. A more sophisticated algorithm could account for variable-sized shakers, which would be useful for environments where there are large differences in specification magnitude between axes.

With respect to test implementation, test setup and control algorithm strategies for low frequency environments could be improved to reduce error associated with shear/moment transfer from stingers. Increasing the minimum frequency cutoff of the environmental specification addresses this error at the cost of misrepresenting the environment. In transportation environments, for example, the low-frequency, high-displacement motion is important to include for qualification tests. Improved experimental strategies could include the selection of flexible stingers to avoid transferring force or active control of shaker position to match the low frequency displacements of the device under test.

This project investigated the use of shaker placement optimization based on FRFs from an updated FEA model in multi-input environmental testing and found that additional considerations, such as shaker impedance and control of low frequency environments, must be considered to make optimization algorithms effective. Our results provide a valuable benchmark for test engineers that use FEA models for multi-input environmental test planning.

Acknowledgments This research was funded by Los Alamos National Laboratory (LANL) through the Engineering Institute's Los Alamos Dynamics Summer School. The Engineering Institute is a research and education collaboration between LANL and the University of California San Diego's Jacobs School of Engineering. This collaboration seeks to promote multidisciplinary engineering research that develops and integrates advanced predictive modeling, novel sensing systems, and new developments in information technology to address LANL mission-relevant problems. The authors would also like to thank Siemens for their donation of their Simcenter Testlab software for use in this project and Thomas Thompson from LANL for his advice on specification development. LA-UR-24-30459.

References

1. *Department of Defense Test Method Standard for Environmental Engineering Considerations and Laboratory Tests* (2000)
2. Hoksbergen, J. "Defining the global error of a multi-axis vibration test". *Sound and Vibration*, 48:8–13 (2014)
3. Daborn, P.M. "Smarter dynamic testing of critical structures" (2014)
4. Pak, C.G. "Multiple shaker placement for ground vibration test of x-59 aircraft using topology optimization". *NASA Armstrong Flight Research Center* (2020)
5. Rohe, D.P., Nelson, G.D., and Schultz, R.A. "Strategies for shaker placement for impedance-matched multi-axis testing". In Walber, C., Walter, P., and Seidlitz, S., editors, *Sensors and Instrumentation, Aircraft/Aerospace, Energy Harvesting & Dynamic Environments Testing, Volume 7*, pages 195–212, Cham (2020) Springer International Publishing.
6. Rohe, D.P., Schultz, R., and Hunter, N. *Rattlesnake User's Manual*. Sandia National Lab. (SNL-NM), Albuquerque, NM (United States), United States (2021)
7. Schultz, R.A. and Nelson, G.D. "Input signal synthesis for open loop multiple-input/multiple-output testing". *Society of Experimental Mechanics* (2020)
8. Natrella, M.G. *Experimental Statistics*. National Bureau of Standards (1963)
9. Cap, J.S. "A tutorial on analysis techniques for deriving mechanical shock and vibration environmental specifications from field data". *Society of Experimental Mechanics* (2021)
10. Lytle, T., Saeger, W., Tombuelt, A., Danforth, S., DeCleck, J., Ouellette, B., and Schultze, J. "Evaluating vibration controller performance in virtual and hardware tests". In *Proceedings of the International Modal Analysis Conference XLII*. Society for Experimental Mechanics (2024)
11. Slater, J. "Examining spatial (grid) convergence" (2021)
12. Rohe, D. "Sdynpy: A structural dynamics python library". In *Proceedings of the International Modal Analysis Conference XLI*. Society for Experimental Mechanics (2023)
13. Rohe, D., Shultz, R., Hunter, N., CERTTECH, and USDOE. *Rattlesnake Vibration Controller v.2.0*. Sandia National Lab. (SNL-NM), Albuquerque, NM (United States), United States (2021)
14. Berens, P. "Circstat: A matlab toolbox for circular statistics". *Journal of Statistical Science*, 31 (2009)

Appendix: Cross Powers from Field Data

In this study, we focused on the auto PSDs for specification development and multi-input test control. However, we also derived specifications for the cross power spectral density, referred to as the "cross power" or "cross PSD" in this appendix, between the two control DOFs. This appendix describes the derivation and result for the interested reader.

In the Environmental Specification section, we computed the auto PSDs S_{xx} and S_{yy} for signals x and y, respectively. We use these quantities to compute the coherence γ_{xy} at each frequency value ω_i ($i \in \{1, 2, \cdots, M\}$ for M frequency lines) for signal block N_j ($j \in \{1, 2, \cdots, P\}$ for P signal blocks):

$$\gamma_{xy}(\omega_i, N_j) = \frac{|S_{xy}(\omega_i, N_j)|^2}{S_{xx}(\omega_i, N_j) * S_{yy}(\omega_i, N_j)}. \tag{17}$$

Using the computed coherence, the magnitude of the cross power $|S_{xy}|$ is given by

$$|S_{xy}(\omega_i, N_j)| = \sqrt{\gamma_{xy}(\omega_i, N_j) S_{xx}(\omega_i) S_{yy}(\omega_i)}. \tag{18}$$

Recall in the Environmental Specification section that we computed the Maximum Predicted Environment (MPE) for the auto PSDs at the control DOFs. To compute the cross power relative to those MPE auto PSDs $S_{xx}^{MPE}(\omega_i)$ and $S_{yy}^{MPE}(\omega_i)$, we can relate the matrix of auto PSD magnitude with cross PSD magnitude by computing conservative "worst case" amplitude estimates for the auto power magnitude and using the definition of coherence. This coherence-defined relationship between auto power and cross power magnitude is important because a coherence greater than 1 is not physically realizable.

$$|S_{xy}(\omega_i)| = \sqrt{\gamma_{xy}(\omega_i, N_j) S_{xx}^{MPE}(\omega_i) S_{yy}^{MPE}(\omega_i)}. \tag{19}$$

This leaves the coherence as the only remaining matrix. Unlike the auto PSD, where the "worst case loading" at each frequency line is a large PSD magnitude, the "worst case" for coherence is unclear. Therefore, we used the mean of the coherence as a reasonable approximation for a cross PSD magnitude that corresponds to the MPE auto PSD:

$$|S_{xy}(\omega_i)| = \sqrt{\overline{\gamma_{xy}}(\omega_i) S_{xx}^{MPE}(\omega_i) S_{yy}^{MPE}(\omega_i)}. \tag{20}$$

Computing the phase of the cross PSD is an additional challenge. For this study, we used the mean phase as computed with the CircStat Matlab toolbox[14]. We computed individual phase values from individual time blocks using Matlab's

CPSD function, then inserted the phase values into the circular statistics mean estimator of CircStat. This process generated a set of phases, $\theta(\omega_i)$, for $i \in \{1, 2, \cdots, M\}$ frequency lines. After the cross PSD magnitude and phase are computed, we recombine them into the complete cross power, given by

$$S_{xy}(\omega_i) = |S_{xy}(\omega_i)| * e^{i\theta(\omega_i)}. \tag{21}$$

Figure 22 shows an example histogram of the cross PSD phase between the two control DOFs at one frequency line.

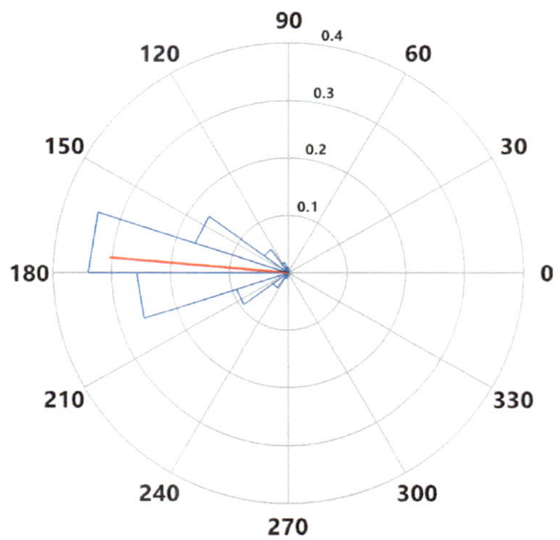

Fig. 22 Histogram for cross PSD phase at 340Hz between DOF 97Z to DOF 58X.

After computing the cross powers between the control DOFs, we assemble the spectral density specification matrix. Figure 23 shows the auto and cross PSDs computed for the two control DOFs in this study.

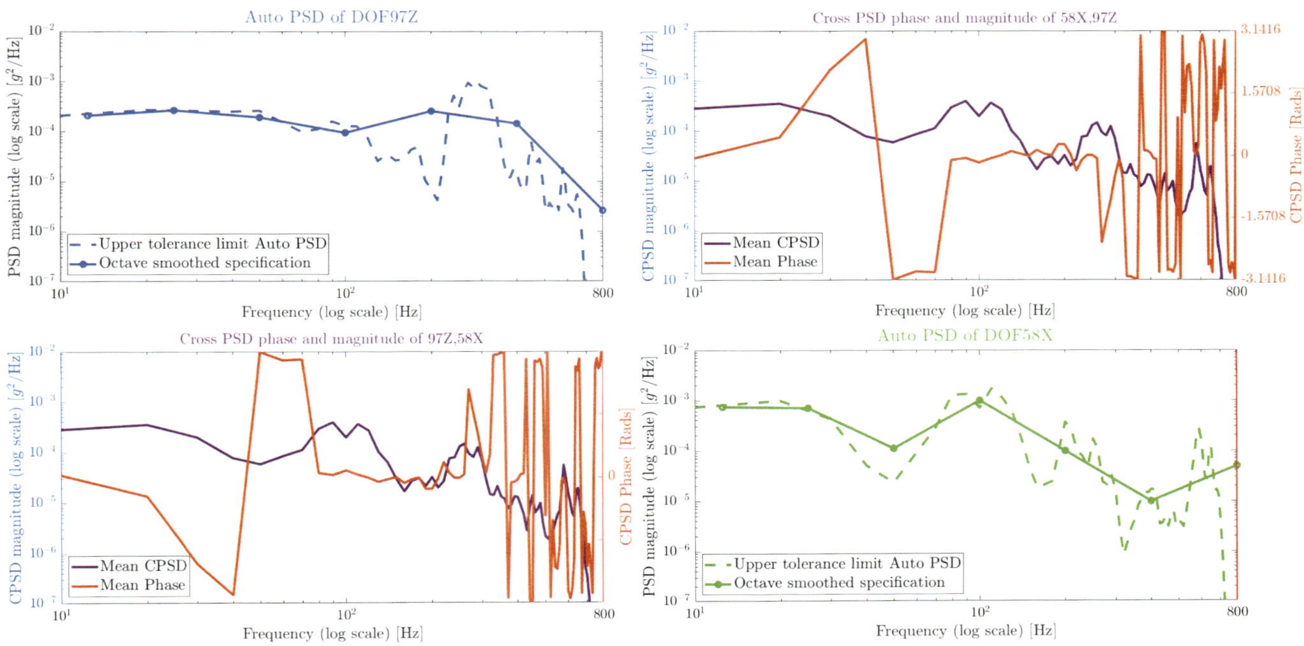

Fig. 23 MPE (P95/50) specification auto and cross PSDs for control DOFs 97Z and 58X.

For Product Safety Concerns and Information please contact our EU representative GPSR@taylorandfrancis.com Taylor & Francis Verlag GmbH, Kaufingerstraße 24, 80331 München, Germany